Laboratory Biosecurity Handbook

Laboratory Biosecurity Handbook

Reynolds M. Salerno
Jennifer Gaudioso

CRC Press
Taylor & Francis Group
Boca Raton London New York

CRC Press is an imprint of the
Taylor & Francis Group, an **informa** business

CRC Press
Taylor & Francis Group
6000 Broken Sound Parkway NW, Suite 300
Boca Raton, FL 33487-2742

Printed in the United States of America on acid-free paper
10 9 8 7 6 5 4 3 2 1

International Standard Book Number-13: 978-0-8493-6475-4 (Hardcover)

Library of Congress Cataloging-in-Publication Data

Salerno, Reynolds M. (Reynolds Mathewson), 1966-
Laboratory biosecurity handbook / Reynolds M. Salerno.
p. cm.
Includes bibliographical references and index.
ISBN 978-0-8493-6475-4 (alk. paper)
1. Biological laboratories--Safety measures--Handbooks, manuals, etc.. 2. Bioterrorism--Safety measures--Handbooks, manuals, etc.. I. Title.

QH323.2.S25 2007
570.28--dc22 2007015712

Visit the Taylor & Francis Web site at
http://www.taylorandfrancis.com

and the CRC Press Web site at
http://www.crcpress.com

Contents

Preface

The intent of this book is to provide guidance to bioscience facilities worldwide on the design and implementation of laboratory biosecurity. Laboratory biosecurity is a combination of systems and practices intended to reduce the risk that legitimate bioscience facilities could be sources of pathogens and toxins for malicious use.

The effective design and implementation of laboratory biosecurity depends on cooperation among various individuals, often from diverse communities, including scientists, technicians, policy makers, security engineers, and law enforcement officials. This book attempts to summarize the concerns that individuals responsible for laboratory biosecurity at a facility will need to consider, as well as the way in which such concerns may overlap or conflict. The book is also intended to provide guidance to the broad international community. Because there is a wide array of bioscience facilities worldwide — some large, some small, some conducting extensive research on dangerous biological agents, some focused only on diagnosis or vaccine development, and others only minimally involved with infectious disease — this book attempts to be broadly applicable by addressing a variety of operating environments and particular challenges to designing and implementing laboratory biosecurity.

Although many of the principles described here are incorporated in some countries' regulatory requirements, this book does not specifically address those regulations. Nor do the recommendations in this book necessarily meet the expectations of specific national regulations. It is the responsibility of the management of individual facilities to ensure that its operations are in compliance with any applicable legal requirements.

The term *laboratory biosecurity* as used in this book and in many other resources[1] refers to a set of systems and practices employed in legitimate bioscience facilities to reduce the risk that dangerous biological agents will be stolen and used maliciously. The term *laboratory biosecurity* has been deliberately selected to reflect the inherent connection with the established practices of laboratory biosafety, which aim to reduce the risk of accidental release of or exposure to infectious disease agents. Together, laboratory biosafety and biosecurity manage the risks of working with these agents in bioscience laboratories, help prevent disease in people, animals, and plants, and minimize the risk of worker injury.

REFERENCES

1. *Biosafety in Microbiological and Biomedical Laboratories*, 4th ed., Washington, D.C., 1999; *Laboratory Biosafety Manual,* 2nd ed. (revised), World Health Organization, Geneva, 2003; Applied Biosafety, *Journal of the American Biological Safety Association*, Web site: http://www.absa.org/resabj.html, accessed 1/23/06;

Tucker, J., Biosecurity: Limiting Terrorist Access to Deadly Pathogens, United States Institute of Peace, 2003, Washington, D.C.; Barletta, M., *Biosecurity Measures for Preventing Bioterrorism,* 2002, http://cns.miis.edu/research/cbw/biosec/pdfs/biosec.pdf; Barletta, M., Sands, A., Tucker, J., Keeping track of anthrax: the case for a biosecurity convention, *Bulletin of the Atomic Scientists,* 58, 03, 57–62.

About the Authors

Both authors are members of Sandia National Laboratories' international biological threat reduction (SNL IBTR) program. SNL is a multidisciplinary technical organization that conducts science and engineering research and development in support of U.S. national security and serves as a technical advisor to the federal government. SNL IBTR works domestically and internationally to prevent and mitigate biological risks caused by highly infectious disease agents. Additional information about the SNL IBTR program can be found at: www.biosecurity.sandia.gov.

Reynolds M. Salerno, Ph.D., is manager of the HSNL IBTR program. His work focuses on international counter-bioterrorism and biological weapons nonproliferation. He and his Sandia team specialize in the safety and security of high-risk pathogens and toxins in laboratories and in transportation systems. Recognized as a leading expert on laboratory biosecurity in the U.S., Salerno and his Sandia team have worked extensively in laboratory biosafety, biosecurity, and disease surveillance and diagnostics internationally. They have visited biocontainment laboratories in dozens of countries specifically to consult on biosafety and biosecurity issues. They have presented major biosafety and biosecurity training workshops in ten different countries since September 2004 and have many more planned in the near future. Salerno has served as a member of the U.S. Delegation to the Biological Weapons Convention (BWC), and in August 2003 presented the U.S. government's approach for securing pathogens at facilities and in transit at the BWC's Experts Group meeting in Geneva. He is a member of the Dual-Use Biological Research Guidelines working group of the U.S. National Science Advisory Board for Biosecurity and the International Criminal and Police Organization's Counter-Bioterrorism Board of Experts. He has been recognized as a Next Generation Fellow on U.S. Global Policy and the Future of International Institutions by the American Assembly at Columbia University. He has published widely in the biological threat reduction and biosecurity field. Salerno earned his Ph.D. degree at Yale University.

Jennifer Gaudioso, Ph.D., is a principal member of the technical staff in the SNLIBTR. Her work focuses on the safety and security of high-risk pathogens and toxins in laboratories and in transportation systems. She specializes in risk assessment, analysis of biological agents, and transport security issues, and is an expert on counter-bioterrorism and biological weapons nonproliferation. She also has extensive laboratory biosafety experience and training, serves on SNL's Institutional Biosafety Committee, and is an active member of the American Biological Safety Association. She has participated in security assessments at

U.S. government bioscience facilities and has contributed to the development of international biosecurity guidelines, conducting biosecurity projects for the U.S. Departments of State, Homeland Security, Energy, Health and Human Services, and Agriculture. Having worked extensively on laboratory biosafety and biosecurity issues internationally, she organized eight international workshops on laboratory biosafety and biosecurity in 2005–2006. Over the past few years, she has consulted on these topics at basic and high containment bioscience laboratories in over ten countries. She earned her Ph.D. degree at Cornell University.

Acknowledgments

This book could not have been written or even attempted without the tremendous hard work, commitment, and support of so many of our mentors, colleagues, friends, and family.

Our work in laboratory biosecurity originally evolved out of Sandia National Laboratories' extensive background and expertise in physical protection. Over the years, we have been mentored by many of Sandia's world renown security experts. We have learned so much about fundamental physical protection issues from the best in the business, including Jim Larson, Bruce Hazlewood, Alan Swanson, Ivan Waddoups, Jim Chapek, John Matter, Basil Steele, and JD Williams.

In our own field, nothing can compare with actual experience in the laboratory designing and implementing biosecurity systems. We have been extremely fortunate to conduct laboratory biosecurity work for many different agencies of the U.S. Government. In particular, we have learned a tremendous amount from our colleagues at the U.S. Army Medical Research Institute of Infectious Diseases, the U.S. Centers for Disease Control and Prevention (CDC), and all of the biocontainment laboratories of the U.S. Department of Agriculture. In addition to this practical experience, our knowledge of laboratory biosecurity issues evolved considerably after assisting the World Health Organization draft international recommendations and helping the CDC and U.S. National Institutes of Health write guidance for the most recent edition of the *Biosafety in Microbiological and Biomedical Laboratories*. The formal relationship that Sandia has established with the American Biological Safety Association has also been intellectually and professionally invaluable. Most recently, the U.S. National Nuclear Security Administration's Office of Nonproliferation Policy and especially the U.S. Department of State's Office of Cooperative Threat Reduction have given us the opportunity to conduct a significant amount of work on laboratory biosecurity with specific countries and overseas laboratories. There have been so many wonderful and gifted people we have interacted with to conduct this broad portfolio of work on laboratory biosecurity, and all of them have significantly influenced our views on this topic.

In terms of direct involvement to this manuscript, the staff of Sandia's International Biological Threat Reduction (IBTR) program made the most significant contributions. Natalie Barnett contributed original drafts of the sections on risk assessment, biosecurity components, and biosecurity recommendations. Lauren Hickok articulated many critical concepts that appear in the program management sections. And John Milloy and George Baldwin are responsible for much of the intellectual basis for the physical protection and material control and accountability sections, respectively. All of the IBTR staff reviewed various sections of this manuscript, and Terri Olascoaga and Robert Huelskamp each provided critical technical suggestions on the entire manuscript. Susan Caskey, Lyle Beck and Kathleen Lowe helped us develop the images and charts for the book. Jason Bolles created the design for the book cover and jacket. And Brenda Chavez helped us revise most of the text. But perhaps the two colleagues from Sandia who contributed most to this book were Heather Mcdaniel, who brilliantly tech-edited the entire manuscript multiple times, always under tight time constraints, and Kathleen Lowe, who served as this book's project manager. Kathleen oversaw every aspect of the creation of this book, and it reflects her true dedication to the IBTR mission. We cannot thank Kathleen enough for all her work on this project.

Sandia National Laboratories provided the financial support for this book. In particular, we would like to express our genuine thanks to Executive Vice President Al Romig, Vice President Les Shephard, and Director Dori Ellis for funding this project. We also sincerely appreciate Madelynne Farber, Patent and Licensing Attorney at Sandia, for her hard work in concluding the contract with CRC Press. We are particularly grateful to Dori Ellis, Terri Olascoaga, and Robert Huelskamp of Sandia's Global Security Programs for their important technical guidance and enduring programmatic support for and faith in the IBTR program.

We also had fantastic assistance and encouragement from our editors at CRC Press. Judith Spiegel, Senior Editor for Life Sciences at the Taylor & Francis Group, originally conceived the idea for this book and has patiently seen it through from start to finish. Project Coordinator Kari Budyk, Project Editor Jim McGovern, and Editorial Assistant Christopher Myron have been wonderful to work with on this project; their collective competence and tolerance is remarkable.

This book would not have been possible without the very significant contributions of all of these great mentors and colleagues, and we cannot overstate our deep gratitude to all of them. Nevertheless, all of the mistakes and shortcomings of this manuscript remain ours and ours alone.

Finally, we extend our deepest gratitude to our respective spouses, Jennifer Salerno and Rob Hemighaus. Each of them has sacrificed by allowing us to dedicate our professional lives to biosecurity and biological threat reduction. Without their support and encouragement, this book would not have been written.

Albuquerque
March 2007

1 Introduction

Bioscience research laboratories, biotechnology institutions, biomedical laboratories, and clinical laboratories are critical in the global fight against emerging and reemerging infectious diseases such as Avian Influenza, SARS, Nipah, Chikungunya, new variant Creutzfeldt–Jakob disease, Lyme borreliosis, epidemic meningitis, Hantavirus, HIV, West Nile, Rift Valley fever, multidrug-resistant *Mycobacterium tuberculosis*, among many others.[1] Every day, bioscience institutions around the world advance humanity's knowledge about infectious disease and develop new technologies to improve the state of public and agricultural health.

Bioscience research, diagnostic, and clinical laboratories also have a critical role to play in countering the threat of bioterrorism. In fact, bioterrorism has been appropriately described as a "deliberately re-emerging infectious disease."[2] These laboratories are on the front lines of emerging and reemerging disease: not only are they called on to respond to outbreaks of infectious disease but their research will pave the way for new vaccines and therapies to counter these diseases. There is little doubt that the advance of bioscience — and the daily work of bioscience institutions — represents the world's best defense against the threat of bioterrorism.

Although the work of these laboratories holds great promise for humanity, the nature of their work also presents safety and security risks that need to be managed. Infectious organisms may infect the workers in laboratories. Harding and Byers define laboratory-associated infections (LAI) as "all infections acquired through laboratory or laboratory-related activities regardless of whether they are symptomatic or asymptomatic in nature."[3] The SARS LAIs in Singapore, Taipei, and Beijing in 2003 and 2004 drew international attention to the issue of laboratory biosafety — those measures taken at a bioscience laboratory to reduce the risk of accidental exposure to or release of a biological agent.

However, many LAIs occur with much less public attention. Although underreporting of LAIs is widely acknowledged,[4] numerous LAIs have been documented in a literature review of the 20-year period from 1980 to 2000:[5] 953 bacterial infections (including rickettsial LAIs), 929 viral infections, 50 parasitic infections, and 5 fungal infections. There were 21 fatalities among these LAIs.

Biosafety, which aims to reduce the risk of LAIs, has been practiced in bioscience laboratories for decades, and there is a body of substantive literature that supports laboratory biosafety activities.[6,7] Clearly, a technical person working with pathogens in a laboratory environment has a vested interest in taking such measures to heart: not doing so could have adverse personal consequences. In addition to protecting occupational workers, laboratory biosafety safeguards the public and the environment from accidents involving pathogens and toxins.

Bioscience laboratory personnel and management have a similar responsibility to implement biosecurity systems, but the concepts of biosecurity are much less understood than biosafety. The literature for laboratory biosecurity is almost nonexistent. This is understandable because laboratory biosecurity has only become a serious issue in the U.S. and throughout the world since the anthrax attacks of 2001.

Laboratory biosecurity recognizes that bioscience facilities and their staffs have an additional stewardship responsibility if they work with biological materials that could harm others should they be used improperly. Laboratory biosecurity helps provide assurance that individuals working with dangerous pathogens and toxins will not deliberately use that material to harm others and will take every reasonable precaution to prevent others from doing so. Whereas biosafety is concerned with *accidental* misuse, biosecurity is concerned with *intentional* misuse.

The third edition of the World Health Organization's (WHO) *Laboratory Biosafety Manual* highlights "the need to protect laboratories and the materials they contain from being intentionally compromised" and introduces the concept of laboratory biosecurity. The WHO recently published a companion document to provide further guidance on laboratory biosecurity. This WHO document on "Biorisk Management" emphasizes that laboratory biosecurity helps protect all valuable biological materials, suggesting that valuable biological materials are not just limited to dangerous pathogens and toxins but extend to scientifically, historically, and economically important biological materials such as collections and reference strains, pathogens and toxins, vaccines and other pharmaceutical products, food products, genetically modified organisms (GMO), nonpathogenic microorganisms, extraterrestrial samples, cellular components, and genetic elements. Thus, the WHO endorses a holistic approach to laboratory biosecurity; it is simply an element of good laboratory practice. Furthermore, WHO writes that "it is the technical and moral duty of laboratory managers and laboratory workers…to reassure the general public…and to prove that the biorisks inherent to laboratory work are controlled with appropriate safeguards to meet their expectations."[8]

1.1 LABORATORY BIOSECURITY AND THE RISKS OF BIOTERRORISM

Experts suggest that the risk of bioterrorism will increase as the technology advances and becomes more widespread and as the costs and technical impediments to creating a biological weapon continue to decline. They predict that, over time, terrorists will resort to bioterrorism instead of attacks with conventional explosives, and that bioterrorism may produce greater consequences than possible today.[9] According to a 2004 report of the National Intelligence Council, "The most worrisome trend has been an intensified search by some terrorist groups to obtain weapons of mass destruction. Our greatest concern is that these groups might acquire biological agents or less likely, a nuclear device, either of which could cause mass casualties ... Terrorist use of biological agents is therefore likely, and the range of options will grow."[10]

Although the use of biological weapons by states in warfare dates back to the Middle Ages, even bioterrorism is not a new phenomenon. A wide range of biological agents have been used by various subnational groups for various reasons over the past century.[11] In 1910, the Pancho Villa guerillas used shards of pottery and obsidian laced with botulinum toxin to attack federal sentries. In 1952, the Kenyan Mau Mau separatist group used African milk brush toxin to poison steer. In 1981, Dark Harvest, an environmental extremist group in the U.K., delivered *Bacillus anthracis*–contaminated soil to a political party conference. The first known bioterrorism incident in the U.S. took place in 1984 when the Rajneeshees, a spiritual movement in Oregon, contaminated salad bars with *Salmonella typhi.*

In April 1990, Aum Shinrikyo, an apocalyptic cult based in Japan, attempted to disseminate botulinum toxin at U.S. Navy bases at Yokohama and Yokosuka, the Narita airport, the Imperial Palace, and the Japanese Diet. Three years later, the group tried to execute bioterrorism attacks six more times in Japan — botulinum toxin on three separate occasions and *Bacillus anthracis* three other times.

In 1995, the year that Aum Shinrikyo disseminated sarin gas on the Tokyo subway, killing 12 people and injuring hundreds, there was a spate of bioterrorism-related incidents. An Afghani warlord in Tajikistan infected Russian troops with the hepatitis virus. Larry Wayne Harris was arrested after receiving three vials of *Yersinia pestis* from the American Type Culture Collection under false pretenses. Diane Thompson tainted donuts at her place of work with *Shigella dysenteriae* type 2 in an attempt to sicken her coworkers, and authorities found a ricin laboratory in the basement of Thomas Leahy's house.

Despite this significant number of bioterrorism-related incidents, culminating with the anthrax attacks in the U.S. that killed 5 people and injured 22, disseminating a biological agent that kills and/or injures peoples or animals is not necessarily a trivial task. The potential bioterrorist must successfully acquire a virulent pathogen or toxin, produce the agent in a suitable form and quantity, and effectively disseminate the agent. Specific knowledge, skills, and equipment are necessary to overcome these technical impediments. Some experts even claim that acquiring a virulent strain is the rate-limiting step.[12]

Experts also suggest that terrorists are likely to use readily available biological agents from laboratories — where they have been characterized and their viability and virulence established—as opposed to isolating agents from nature or pursuing the creation of novel pathogens.[13] Although nearly all dangerous biological agents can be isolated from nature, successfully obtaining a viable and virulent strain of a dangerous pathogen from a natural source is a highly technical, difficult, and time-intensive undertaking.

In theory, terrorists have the option of constructing viable and virulent organisms synthetically or through genetic engineering of a relatively harmless agent. Yet, today, there is a relatively low risk that biological agents will be created by *de novo* synthesis or genetic engineering specifically for bioterrorism purposes. Terrorist groups would have to allocate extensive resources for a research and characterization-testing program based on the creation of a new organism. Moreover, the technical challenges to such an approach are substantial. Although it is possible to engineer a harmless agent to display heightened virulence or to achieve antibiotic resistance in an existing agent, producing such a transformation is technically challenging, time-consuming, and costly.

The existence of dangerous biological agents in natural environments and the availability of genetic engineering and chemical synthesis technology demonstrate that laboratory biosecurity will never definitively prevent bioterrorism. A terrorist need not steal an agent from a bioscience laboratory to disseminate an infectious disease maliciously. Moreover, the nature of biological materials and the manner in which bioscience is conducted — issues that will be explored in more detail in this book — make achieving highly reliable laboratory biosecurity almost impossible. No security system can be developed that definitively prevents a determined individual from removing a biological agent from a laboratory undetected.

Nevertheless, it is now widely acknowledged that managing dangerous pathogens and toxins in legitimate bioscience facilities is a critical bioterrorism prevention strategy. Laboratory biosecurity — implemented in a way that does not compromise critical research and diagnostics — mitigates the risk that a legitimate bioscience facility could become a source of material for malicious use.

1.2 LABORATORY BIOSECURITY AND INTERNATIONAL OBLIGATIONS

It is appropriate that most national strategies to counter the threat of bioterrorism focus on improving the ability of a country to detect and respond to the next bioterror attack. However, recently, many countries have also emphasized the need for preventive measures, such as laboratory biosecurity. The U.S. Homeland Security Presidential Directive 10 (HSPD-10), *Biodefense for the 21st Century*, defines "Prevention and Protection" as one of the four essential pillars of the U.S. response to the bioterrorist threat: "Preventing biological weapons attacks is by far the most effective approach to biodefense. Prevention requires the continuation and expansion of multilateral initiatives to limit the access of agents, technology, and know-how to countries, groups, or individuals seeking to develop, produce, and use these agents."[14]

The U.S. scientific community has also recognized the national security challenges that the expanding biotechnology industry is creating. In 2003, the National Academy of Sciences published "Biotechnology Research in an Age of Terrorism," a report imploring the scientific and policy communities to take action in curbing the biological weapon (BW) proliferation opportunities brought about by the rise of the global biotechnology industry.[15] Largely as a result of this study, the U.S. government established the National Science Advisory Board for Biosecurity (NSABB), which is responsible for considering codes of conduct for the bioscience community and developing a system of institutional and federal review of bioscience research.[16] Most recently, the National Academy published *Globalization, Biosecurity, and the Future of the Life Sciences*, which concludes that the broad spectrum of biological threats will continue to expand in the future and implores the "global community of life scientists to adopt a common culture of awareness and a shared sense of responsibility" to mitigate these security concerns.[17]

The international community has also recognized the need to develop preventive solutions to counter the contemporary bioterrorist threat. The International Committee of the Red Cross (ICRC) recently called on governments to strengthen their commitment to international legal norms that prohibit the hostile uses of biological agents, and safeguard potentially dangerous agents, technology, and knowledge.[18] In 2005, the International Criminal Police Organization (Interpol) launched an initiative to enhance the controls on and prevention of the malicious use of pathogens and toxins.[19]

Beyond these awareness-raising initiatives, the international community has also created a set of international obligations on laboratory biosecurity. The Bacteriological (Biological) and Toxin Weapons Convention (BWC) addresses aspects of biosecurity throughout its text.[20] Article IV requires each state party to enact implementing legislation for the enforcement of the BWC, placing an obligation on states to control misuse of dangerous biological agents by both state agencies and non-state agencies within its jurisdiction or control. Multiple articles address the concept of national pathogen security. Article II mandates state parties to take appropriate measures to protect the public and the environment from dangerous biological agents. Article III indicates that state parties cannot provide resources to others to misuse biological agents, and Article VII compels state parties to assist a state party that has been harmed by the misuse of biological agents. In essence, the BWC asserts that state parties are responsible for the safety and security of their biological resources and must attempt to ensure that other state parties cannot be harmed by biological weapons. At the BWC's technical experts meeting in 2003, state parties agreed to pursue national implementation of laboratory and transportation biosecurity. They renewed this commitment at the 2006 BWC Review Conference.

More recently, United Nations Security Council Resolution (UNSCR) 1540 urged states to take preventive measures to mitigate the threat of biological, chemical, and nuclear terrorism.[21] UNSCR 1540 requires states to establish and enforce legal barriers to prevent the acquisition of weapons of mass destruction (WMD) by terrorists and states. Paragraph 3 of the resolution mandates that states "Take and enforce effective measures to establish domestic controls to prevent the proliferation of...biological weapons...including by establishing appropriate controls over related materials." Specifically, states need to develop and maintain appropriate effective measures to account for and secure such items in production, use, storage, or transport, and they need to develop and maintain appropriate effective physical protection measures. States are obligated to submit reports to the 1540 Committee that document their efforts to meet the expectations of the United Nations (U.N.) resolution.

Implementation of laboratory biosecurity helps nations comply with the international obligations enshrined in the BWC and UNSCR 1540. Moreover, laboratory biosecurity demonstrates a nation's commitment to the global norm against the misuse of dangerous biological agents.

1.3 LABORATORY BIOSECURITY AND NATIONAL REGULATIONS

To meet these international obligations, countries have implemented a series of national measures: BWC-implementing legislation that bans the development, production, and stockpiling of biological weapons; national legislation and regulations for laboratory and transportation biosecurity to control the use, storage, and transport of dangerous pathogens and toxins domestically; export control regulations to control the transport of dangerous biological agents and technology internationally; and implementation standards or guidelines to provide assistance to those who handle, store, or transport dangerous biological agents. Implementation guidance is critical to ensure that laboratories can comply with legislation while still meeting their biomedical and bioscience research and diagnostic responsibilities.

This section provides a short introduction to a variety of national implementation measures to illustrate differences and commonalities in approaches. The information in this section is by no means a comprehensive compilation of national implementation measures; it is simply a summary of some significant national initiatives. The Web sites of Interpol[22] and the Organisation for Economic Co-operation and Development (OECD)[23] were sources for this information.

Canada passed its "Biological and Toxin Weapons Convention Implementation Act" in 2004. This act made it illegal to possess biological material except for peaceful purposes and illegal to possess any weapon, equipment, or other means of using biological materials for hostile purposes. Within the Public Health Agency of Canada, the Office of Laboratory Security has responsibility for regulating biosecurity and certifying containment laboratories. That office publishes Laboratory Biosafety Guidelines that address both laboratory biosafety and laboratory biosecurity.[24]

India's "Prevention of Terrorism Act" of 2002 bans possession of biological substances for use in warfare; this is the BWC-implementing legislation for India. India also has several national regulations that address bioterrorism. The "Weapons of Mass Destruction and Their Delivery Systems (Prohibition of Unlawful Activities) Bill" prohibits all unlawful activity related to WMD, including their means of delivery and dual-use materials, equipment, and technology. The "Rules for the Manufacture, Use, Import, Export, and Storage of Hazardous Microorganisms, Genetically-Engineered Organisms, or Cells," which is administered by the Department of Biotechnology in the Ministry of Science and Technology, regulates the use, import, and export of listed biological agents and limits research with regulated biological agents to government-approved laboratories.

The BWC-implementing legislation in Japan went into effect in 1982. In October 2001, the Ministry of Health, Labour, and Welfare issued a "Ministerial Notice on Laboratory Safeguards," which requested that laboratories establish safeguard systems for dangerous pathogens, ensure adequate security measures, keep a register of dangerous pathogens, and maintain records of the possession of select agents as well as their stated purposes.

The Malaysians cover their obligations for BWC-implementing legislation under a multitude of separate legislative documents, including the Penal Code, the Corrosive and Explosive Substances and Offensive Weapons Act of 1958, the Arms Act of 1960, the Customs Act of 1967, the Plant Quarantine Act of 1976, and the Prevention and Control of Infectious Diseases Act of 1988. In 2005, Malaysia established a National Biosafety Committee and tasked it with writing a "Biological Agents and Toxins Act" and a "National Laboratory Biosafety and Biosecurity Policy." Implementation standards will be developed after the completion of these national policy documents.

Singapore has strict regulatory requirements for laboratory biosafety and biosecurity, implemented in the "Biological Agents and Toxins Act" of 2005. This act provides the framework for regulating biological activities in Singapore, addressing possession, use, transfer, import, export, transport, and biosafety. It establishes penalties of up to a $1 million dollar fine and life imprisonment for violation of the act. The WHO *Laboratory Biosafety Manual* has been adopted as the Singapore standard for laboratory biosafety.

The U.K. has a robust regulatory framework for laboratory biosecurity; an overview of only a few of the regulations is provided here. The U.K. enacted the "Biological Weapons Act" in 1974. The "Anti-Terrorism, Crime, and Security Act of 2001" has multiple parts that address bioterrorism — Part 6 amends the "Biological Weapons Act" to address the transfer of biological materials, and Part 7 requires security of pathogens and toxins. The implementation of Part 7 requires close coordination of laboratories and local police. The "Control of Substances Hazardous to Health" governs laboratory-related exposures, among other topics. The U.K. has also created an Advisory Committee on Dangerous Pathogens to develop implementation guidance and standards, including notification of use requirements for certain biological agents.

The 1989 "Biological Weapons Anti-Terrorism Act" serves as the BWC-implementing legislation for the U.S. The current U.S. biosecurity regulatory environment is based on the "USA PATRIOT Act" and the "Public Health Security and Bioterrorism Preparedness and Response Act of 2002," which aim to improve the protection of "select" agents and toxins. Three Codes of Federal Regulations (42 CFR 73, 7 CFR 331, and 9 CFR 121) implement those acts by establishing lists of biological agents that pose a threat to humans, animals, or plants. These regulations require any laboratory that possesses one of these pathogens or toxins to enforce and adhere to specific security measures. The U.S. Department of Transportation (DoT) has requirements for biosecurity of these select agents during transport. The *Biosafety in Microbiological and Biomedical Laboratories*[6] (*BMBL*), published by the U.S. Centers for Disease Control and Prevention (CDC) and the U.S. National Institutes of Health (NIH), provides guidance on laboratory biosafety; the fifth edition of these guidelines discusses laboratory biosecurity more substantively way than the fourth edition.

Overall, countries rely on a mixture of guidance and regulation for implementing laboratory biosafety and biosecurity, and there are many different models for oversight of bioscience laboratories. Many countries are just now beginning to address these issues at a national level.

1.4 APPROACH/OBJECTIVES OF THIS BOOK

The basic principles of laboratory biosecurity described in this book are a reflection of the concepts that are becoming enshrined in national regulations of multiple countries — a determination of which biological materials require what accountability; and reliability of personnel with access; limiting access to dangerous biological agents; and basic principles of physical security.

There are two fundamental questions that the management of a bioscience facility needs to answer before it can establish an effective biosecurity system: What things of value — or assets — will the biosecurity system protect and what threats will the biosecurity system protect those assets against? In general, the more assets a security system must protect, and the more threats the security system must protect those assets against, the more intrusive the security system will be to the facility's operations and the more expensive it will be to install and maintain.

For the purposes of this book, laboratory biosecurity will focus on preventing the theft and misuse of biological agents from a bioscience laboratory. A biosecurity system should not necessarily be limited to protecting pathogens and toxins from theft — and the methodology presented here can be extended to other assets and other malicious actions — but every laboratory biosecurity system should at least include this objective.

Designing and implementing a cost-effective laboratory biosecurity system depends, first and foremost, on recognizing the unique nature of biological assets.[25] Traditional security approaches applied to very high-value items or unique dangerous materials, such as special nuclear materials, may not directly translate to the operating reality of the bioscience institution. As the earlier short review of bioterrorism incidents indicated, many different pathogens and toxins can be used maliciously. With few exceptions, biological agents occur naturally. Dangerous pathogens and toxins are also worked with at hundreds of legitimate bioscience laboratories around the world. It is important for security designers to understand that a legitimate bioscience institution's possession of a dangerous biological agent does not imply malicious intent. All dangerous biological materials are known as fundamentally "dual-use" materials: They have legitimate defensive and peaceful commercial, medical, and research applications, but they could also be misused to intentionally cause infectious disease.

Pathogens are self-replicating organisms, making the theft of any amount potentially significant for a bioterrorist. The notion of self-replication, as well as unpredictable mutation and death, also prohibits a reliable quantitative accounting of biological agents through their life cycle. At a bioscience institution, biological agents can be found in many different physical locations in many different forms. For example, biological agents exist in freezers, incubators, animals, carcasses, animal bedding, and waste material. This decentralization makes restricting access to authorized individuals extremely difficult. Also, in contrast to many nuclear materials chemicals, biological agents emit so little energy that they are not detectable with "stand-off sensors," making it virtually impossible for a security system to identify someone illicitly removing biological materials from a facility.

Security system designers must be cognizant of these challenges to protecting biological agents if they are to design a laboratory biosecurity system that does not jeopardize the operations at the bioscience institution. The wide range of bioscience institutions and types of work with biological agents suggests that there is not a one-size-fits-all biosecurity solution. Many bioscience institutions have pathogens and toxins and, as a result, should implement prudent laboratory biosecurity measures. However, not every bioscience facility houses collections of dangerous pathogens and toxins. Rather, there are a limited number of facilities around the world that work with, characterize, and store those biological agents at the highest risk of terrorist misuse. It is only these few facilities that need to implement substantive biosecurity.

Implementing laboratory biosecurity should begin with a risk assessment, which defines and prioritizes the biosecurity risks at a facility and serves as the basis for establishing the specific objectives of the biosecurity system. Chapter 2 describes the biosecurity risk assessment process. There are five principal components of a laboratory biosecurity system — physical security, personnel security, material control and accountability (MC&A), transport security, and information security — that can be implemented in various degrees to achieve the objectives articulated through the risk assessment process. Chapter 3 describes these five components of laboratory biosecurity in detail. Coherent and consistent management and oversight of the laboratory biosecurity system is essential to ensure that it operates effectively within the operating constraints of the institution. Chapter 4 focuses on the program management elements of a laboratory biosecurity system. Chapter 5 provides specific laboratory biosecurity recommendations based on our experience at bioscience institutions around the world. The appendices are intended to serve as resources for biosecurity system designers.

In writing this book, we hope to provide a logical framework for a wide range of bioscience institutions that have a responsibility to protect their dangerous pathogens and toxins against theft and misuse. Because specific national regulations or requirements for laboratory biosecurity vary widely, the recommendations in this book may or may not necessarily constitute adequate guidance for assisting a facility in meeting its own legal obligations. However, based on our international experience, we believe that many, if not all, of the laboratory biosecurity concepts presented here are applicable to bioscience facilities around the world.

Specifically, this book is an attempt to begin documenting some of the methods and procedures for designing and implementing a laboratory biosecurity system. As with biosafety, laboratory biosecurity is a subjective exercise that can be achieved successfully in a variety of ways. This book is not intended to describe the only way in which to develop a laboratory biosecurity system; it is simply an explanation of one process that has been used successfully in a wide range of bioscience facilities around the world. Moreover, the concept of laboratory biosecurity is in its infancy. Many methods, procedures, and technologies are still evolving or under initial development. As such, the ideas expressed in this book are largely conceptual and subject to much review and critique.

REFERENCES

1. Fauci, A.S., 2006, Robert H. Ebert Memorial Lecture: Emerging and Re-Emerging Infectious Diseases: The Perpetual Challenge, New York: Milbank Memorial Fund.
2. Fauci, A.S., Morens, D.M., and Folkers, G.K., 2004, The challenge of emerging and re-emerging infectious diseases, *Nature,* 430, 242–249.
3. Byers, K.B. and Harding, A.L., 2000, Epidemiology of laboratory-associated infections, in *Biological Safety: Principles and Practices*, Fleming, D.O. and Hunt, D.L., Eds., Washington, D.C.: ASM Press, pp. 35–54.
4. Sewell, D.L., 1995, Laboratory-associated infections and biosafety, *Clinical Microbiology Reviews,* 8(3), 389–405.
5. Byers, K.B. and Harding, A.L., 2000, Epidemiology of laboratory-associated infections, in *Biological Safety: Principles and Practices*, Fleming, D.O. and Hunt, D.L., Eds., Washington, D.C.: ASM Press, pp. 35–54.
6. U.S. Department of Health and Human Services, Centers for Disease Control and Prevention, and National Institutes of Health, 1999, *Biosafety in Microbiological and Biomedical Laboratories (BMBL)*, 4th ed., Washington, D.C.: publisher?
7. World Health Organization, 2004, *Laboratory Biosafety Manual*, 3rd ed., Geneva: WHO.
8. World Health Organization, 2006, Biorisk Management: Laboratory Biosecurity Guidance, 1st ed., Geneva: WHO.
9. National Academy of Sciences, 2003, *Biotechnology Research in an Age of Terrorism: Confronting the Dual Use Dilemma*, Washington, D.C.: National Academy Press; Nixdorff, K. and Bender, W., March 2002, Biotechnology, Ethics of Research, and Potential Military Spin-off. *INESAP Bulletin*, 19, http://www.inesap.org/bulletin19/bul19art05.htm; Kwik, G. et al., 2003, Pathogen security: responsible stewardship in an age of catastrophic terrorism, *Biosecurity and Bioterrorism,* 1(1); Salerno, R.M. et al., 2004, A biological weapons risk assessment based on historical and technical perspectives, *The Nonproliferation Review,* 11(3); Central Intelligence Agency Office of Transnational Issues, 2003, The Darker Bioweapons Future, Washington, D.C.: CIA.
10. National Intelligence Council, 2004, *Mapping the Global Future*, Pittsburgh, PA: NIC.
11. Salerno, R.M. et al., 2004, A biological weapons risk assessment based on historical and technical perspectives, *The Nonproliferation Review,* 11(3), 24–55.
12. Smithson, A.E. and Levy, L.-A., 1999, Stimson Center Report No. 35, Ataxia: The Chemical and Biological Terrorism Threat and the U.S. Response, Washington, D.C.: Stimson Center.

13. Scheuer, M., 2002, *Through Our Enemies' Eyes: Osama Bin Laden, Radical Islam, and the Future of America*, Dulles, VA: Brassey's; Petro, J.B. and Relman, D.A., 2003, Understanding threats to scientific openness, *Science,* 302, 1898; Office of the Press Secretary, The White House, 2004, Commission on the Intelligence Capabilities of the United States Regarding Weapons of Mass Destruction: Report to the President of the United States, Washington, D.C.: White House.
14. U.S. Homeland Security Presidential Directive 10 (HSPD-10), Also known as National Security Presidential Directive 33 (NSPD-33), 2004, Biodefense for the 21st Century, Washington, D.C.: White House.
15. National Academy of Sciences, 2003, Biotechnology Research in an Age of Terrorism: Confronting the Dual Use Dilemma, Washington, D.C.: National Academy Press.
16. Office of Biotechnology Activities, National Institutes of Health, National Science Advisory Board for Biosecurity, 2005, http://www.biosecurityboard.gov/index.asp.
17. National Academy of Sciences, 2006, *Globalization, Biosecurity, and the Future of the Life Sciences*, Washington, D.C.: National Academy Press.
18. International Committee of the Red Cross, 2002, Biotechnology, Weapons and Humanity: Appeal of the International Committee of the Red Cross, Geneva: ICRC.
19. Interpol, 2005, Preventing Bioterrorism: The 1st Interpol Global Conference to Strengthen Law Enforcement Preparedness and Develop Effective Police Training, http://www.interpol.int/Public/BioTerrorism/Conferences/1stGlobalConference.asp.
20. The Biological and Toxin Weapons Convention, http://www.opbw.org/.
21. U.N. Security Council Resolution 1540, 2004, New York: United Nations, http://www.state.gov/documents/organization/32097.pdf.
22. http://www.interpol.int/Public/BioTerrorism/NationalLaws/.
23. http://www.biosecuritycodes.org/leg.htm.
24. http://www.phac-aspc.gc.ca/ols-bsl/lbg-ldmbl/index.html.
25. Tucker, J.B., 2003, Biosecurity: Limiting Terrorist Access to Deadly Pathogens, United States Institute of Peace, *Peaceworks* 52.

2 Risk Assessment

Although the term *risk* is common vernacular, it is rarely acknowledged that risk is a function of probability and consequences. In other words, higher risk can be a result of an increase in probability that an event will take place or an increase in the consequences if the event were to occur. In order to answer the question "What is the risk?" three subsidiary questions must be answered: "What can go wrong?" "How likely is it?" "What are the consequences?"[1] A risk assessment is a systematic, structured process for analyzing and determining risk, and the risk assessment should serve as the basis for managing risk or deciding how best to mitigate risk.

Although this chapter focuses on risk assessment for laboratory biosecurity, the same general process is applicable to the design and implementation of a laboratory's biosafety system. A biosafety risk assessment determines the appropriate biosafety precautions that need to be taken to mitigate the risk of accidental release of or exposure to a biological agent during a particular laboratory procedure.

Because the objectives of laboratory biosafety and laboratory biosecurity are different, the definition of risk will be different for these two risk assessments. For a biosafety risk assessment, risk varies with

1. The likelihood of release of or exposure to a biological agent during a specific laboratory procedure
2. The severity of the consequences of that release or exposure on laboratory workers and/or the environment

For a biosecurity risk assessment, risk varies with

1. The likelihood of theft of a biological agent
2. The severity of the consequences of an attack with that agent

The potential consequences of a bioterrorism event with a particular agent generally can be assessed with available scientific and technical data. However, it is impossible to quantitatively determine the probability of an individual deliberately targeting a facility in order to steal and misuse a biological agent. Instead, we evaluate the likelihood or potential that such an event could occur; we use the term *threat potential* to describe this element of biosecurity risk (Figure 2.1). Evaluating the threat potential and severity of the consequences of the theft and misuse of a biological agent is the essential first step for facility managers in developing a coherent laboratory biosecurity program.

$$\text{Biosecurity Risk} = \left(\text{Threat Potential}\right) * \left(\text{Consequences}\right)$$

FIGURE 2.1 Basic Biosecurity Risk Equation.

Understanding the risks and the uncertainties involved is critical for management to responsibly allocate its resources. The allocation of resources and the implementation of operational restrictions should be at the discretion of facility management, but the application should always be in a graded manner — protecting the assets at the highest risk more than those at lower risks. The U.S Government Accountability Office (GAO) has endorsed a risk-management approach to security, which (1) establishes which assets should be protected against which threats, and (2) ensures that the amount of protection provided to a specific asset, and the cost for that protection, is proportional to the risk of the theft or destruction of that asset.[2]

Fundamentally, a security risk assessment is an evaluation and prioritization of a series of potential undesired events — or scenarios — that could occur at a facility. Each scenario is a combination of three components:

1. A specific asset, such as a pathogen or toxin
2. An individual or group of individuals who wish to cause harm
3. A particular action, such as the theft of a pathogen or toxin

There are many different items of value — or assets — at bioscience facilities. When asked, managers often describe their employees, their facilities, and their equipment as their most valuable assets. However, certain information and certain materials, such as pathogens and toxins, are also important assets to consider protecting. Even the types of agents stored and used at a bioscience facility vary widely, and the vast majority of biological agents are not good candidates for use in bioterrorism. Thus, a laboratory's assets should be evaluated individually within the risk assessment process.

In commonly used security terminology, the individuals who intend to attack a facility — either overtly or covertly — to obtain an item illegally or to cause harm are known as *adversaries*. The purpose of a biosecurity system is to mitigate the risk of an adversary successfully exploiting a legitimate bioscience laboratory so that he or she may more easily commit bioterrorism. Because there are a variety of people who might attempt to use the resources of a bioscience laboratory maliciously, the risk assessment should also evaluate each of these adversaries individually.

Facility security systems can be designed to protect against many different undesired events. For instance, a security system can protect items of value against theft. These assets could include equipment, information, or specific materials. A security system can also protect certain assets against destruction or sabotage. This might include vehicles, laboratories, or specific infrastructure, such as electrical substations or decontamination facilities. A security system might also be designed to protect individuals who work at a facility against harm. The risk assessment should also evaluate each specific possible action that an adversary could take at a facility individually, such as theft, sabotage, or personal assault.

As indicated in the Introduction, this book is not intended to provide guidance for every possible security consideration that a bioscience laboratory might face. This book is focused exclusively on providing conceptual guidance for designing and implementing a biosecurity system that aims to prevent the theft and misuse of biological agents from a bioscience laboratory. Although a biosecurity system should not necessarily be limited to protecting pathogens and toxins from theft (the methodology presented here can be extended to other assets and other malicious actions), every biosecurity system should at least include this objective. Thus, the biosecurity risk assessment process described in this chapter will concentrate on evaluating the threat potential and consequences of specific scenarios, which include these three components:

1. A specific pathogen or toxin
2. An individual or group of individuals who wish to steal a pathogen or toxin from a bioscience laboratory
3. The theft of a pathogen or toxin

Because the risk of intentional theft and misuse of pathogens from a bioscience laboratory cannot be eliminated without ceasing all work with pathogens, management has an obligation to distinguish between "unacceptable" and "acceptable" risks at the laboratory. The biosecurity risk assessment process prioritizes biosecurity scenarios by risk, allowing management to determine and decide which scenarios represent unacceptable risks and which scenarios represent acceptable risks. Security designers can then identify appropriate and cost-effective countermeasures to mitigate the unacceptable risks.

In other words, risk assessment provides the fundamental basis for practicing responsible risk management.[3] As the National Research Council's Committee on Risk Characterization has stated, the risk assessment process "must (1) be decision driven, (2) recognize all significant concerns, (3) reflect both analysis and deliberation, with appropriate input from the interested and affected parties, (4) be appropriate to the decision."[4] What follows is a conceptual discussion of the critical components for assessing laboratory biosecurity risk.

2.1 OVERVIEW OF BIOSECURITY RISK ASSESSMENT METHODOLOGY

Laboratory biosecurity risk management is a combination of risk assessment and risk reduction. Risk assessment determines and prioritizes the risks that exist at a laboratory. Risk reduction establishes how those risks will be specifically mitigated. The risk assessment process for a bioscience facility can be summarized in this way:[5]

1. Characterize assets and threats.
   a. Evaluate the pathogens and toxins at the facility (asset assessment).
   b. Evaluate the adversaries who might attempt to steal those pathogens or toxins (threat assessment).
2. Evaluate scenarios.
   a. Create scenarios that consist of a specific adversary attempting to steal and misuse a specific biological agent.
   b. Determine how the various scenarios could be perpetrated.
3. Characterize the risk.
   a. Evaluate threat potential and consequences of each scenario.
   b. Determine acceptable and unacceptable risks; develop risk statement or definition.

The problem should be defined in terms of criteria that are relevant to the problem, understandable, measurable, and nonredundant.[6] One specific implementation of this methodology is described in Appendix B.

2.2 CHARACTERIZE ASSETS AND THREATS

Assessing the assets and threats for laboratory biosecurity consists of:

1. Evaluating the pathogens and toxins at the facility
2. Evaluating the adversaries who might attempt to steal those pathogens or toxins

2.2.1 Evaluate the Pathogens and Toxins

Evaluating the biosecurity risk that pathogens and toxins present at a facility is the first step in the biosecurity risk assessment process. This evaluation has two principal components:

1. Assessing the biochemical properties of the pathogens and toxins to determine how easy or difficult it would be to successfully use them maliciously ("agent task complexity")
2. Assessing the potential consequences of malicious use of those pathogens and toxins

2.2.1.1 Ease or Difficulty of Malicious Use

Most bioscientists are familiar with conducting agent-specific safety risk assessments. For a biosafety risk assessment, the first step is to review what is known about the agent, including whether the agent is associated with LAIs, toxicity, oncogenicity, or allergies. This review enables the agent to be assigned to a safety risk group.

The third edition of the WHO's *Laboratory Biosafety Manual* describes biosafety risk groups as the starting point for determining appropriate biosafety

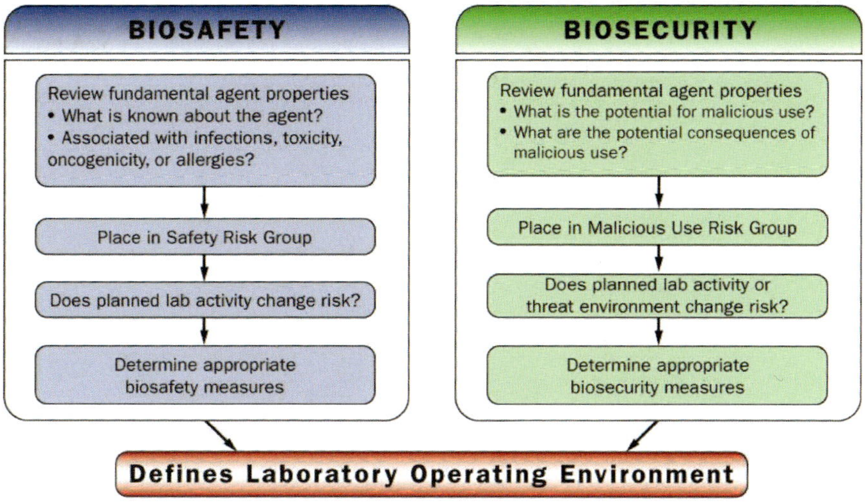

FIGURE 2.2 Parallels between biosafety and biosecurity risk assessment methodologies.

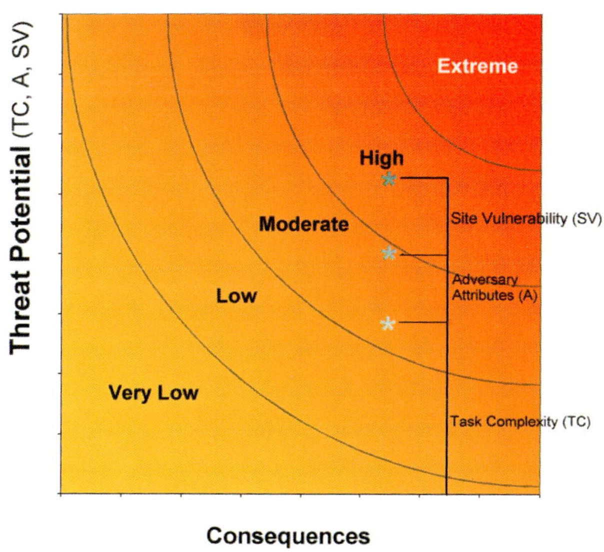

FIGURE 2.3 Elements of biosecurity risk.

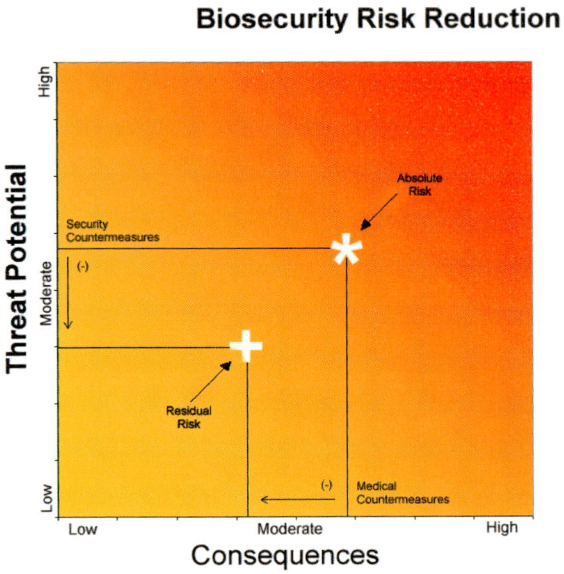

FIGURE 2.5 Risk reduction.

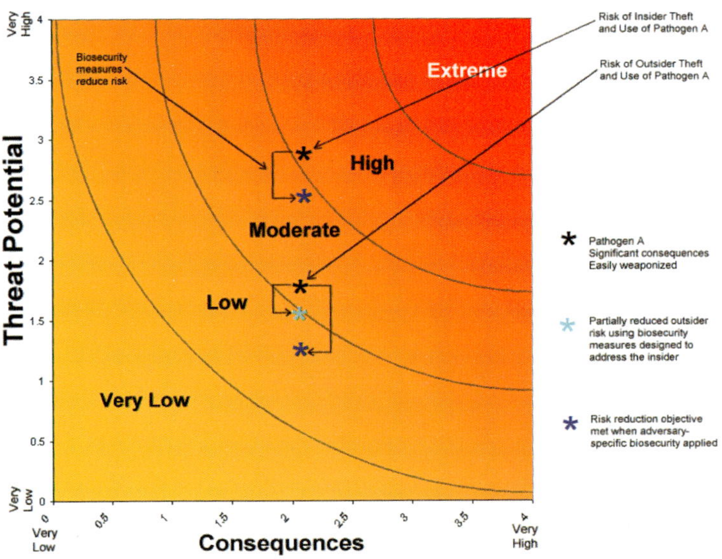

FIGURE 2.6 Insider vs. outsider threat reduction.

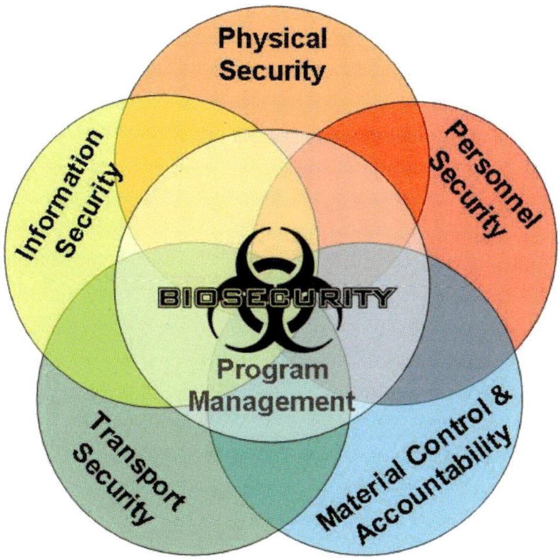

FIGURE 3.1 Components of biosecurity.

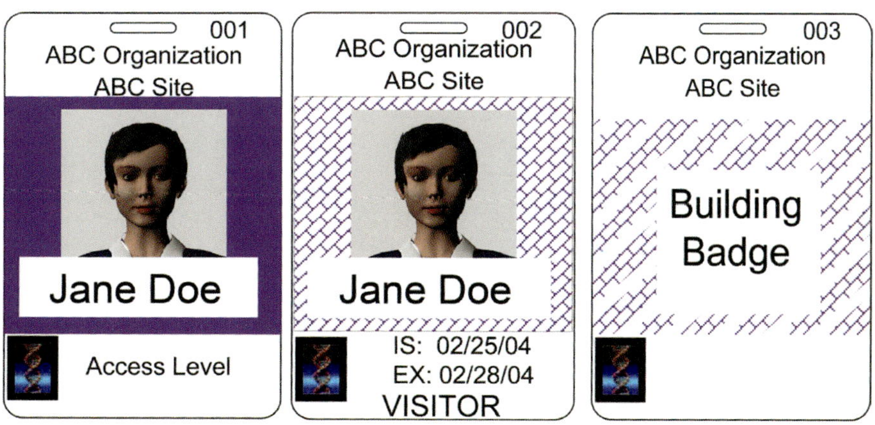

FIGURE 3.2 Identification badges.

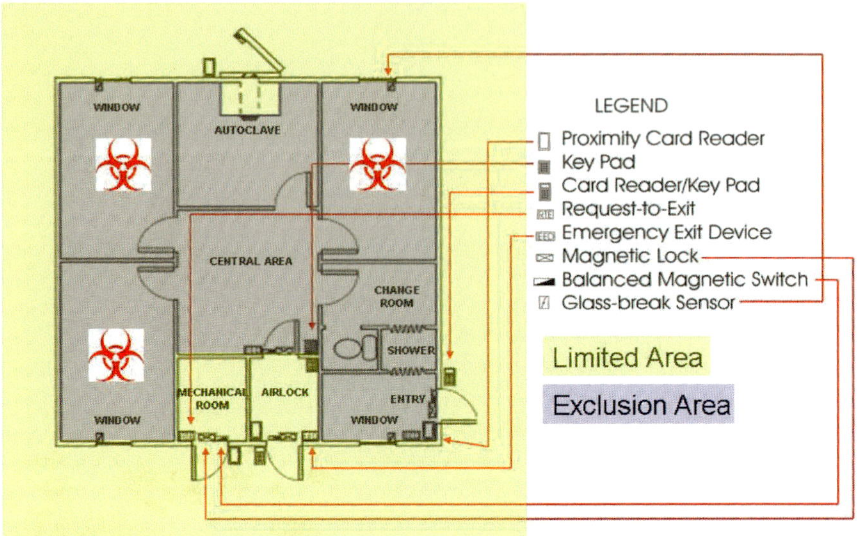

FIGURE 4.4 Example of a physical security layout.

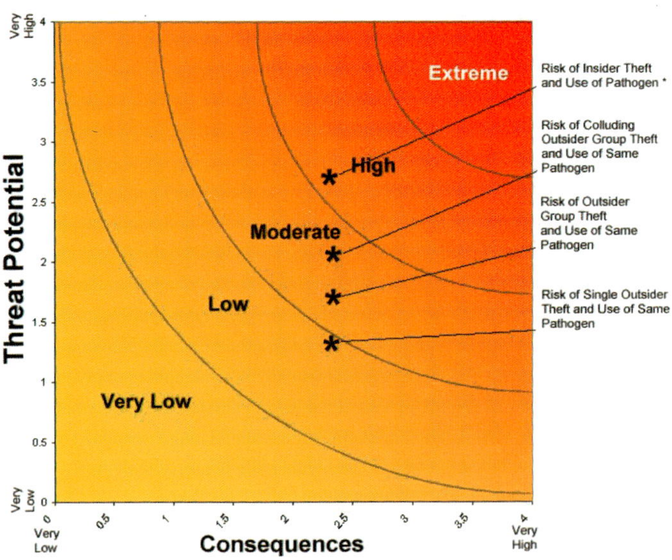

FIGURE B.1 Hypothetical risk assessment results for one pathogen.

measures: "One of the most helpful tools available for performing a microbiological risk assessment is the listing of risk groups for microbiological agents. However, simple reference to the risk grouping for a particular agent is insufficient in conducting a risk assessment. The assignment of a biosafety level takes into consideration the organism (pathogenic agent) used, the facilities available, and the equipment practices and procedures required to conduct work safely in the laboratory."[7]

A biosecurity risk assessment should start with an analogous review of the agent's fundamental properties,[8] but then should focus on whether those biochemical properties of the agent make it an effective weapon. The maximum credible potential consequences of the agent and the effects that it would have on the population at large should be considered. An agent's characteristics influence the potential consequences of malicious use of that agent, such as the numbers of people, animals, or plants killed or sickened, as well as economic and social impacts. The parallels between biosafety risk assessment and biosecurity risk assessment are summarized in Figure 2.2.

An analysis of an agent's potential for malicious use should include factors such as the availability of a suitable strain, ease of production of an appropriate quantity in an appropriate form, modes of dissemination, stability of the agent both in the laboratory and after dissemination, and the availability and level of knowledge required to use the agent as a weapon. We describe this as the "agent task complexity" associated with malicious use. Agent task complexity may be viewed as the difficulty of acquiring, producing, and disseminating a biological agent maliciously. Specifically, agent task complexity is a function of three criteria:

1. Difficulty of acquiring the agent (acquisition)
2. Difficulty of processing the agent into a suitable quantity in a suitable form (development)
3. Difficulty of disseminating the agent to cause harm (dissemination)

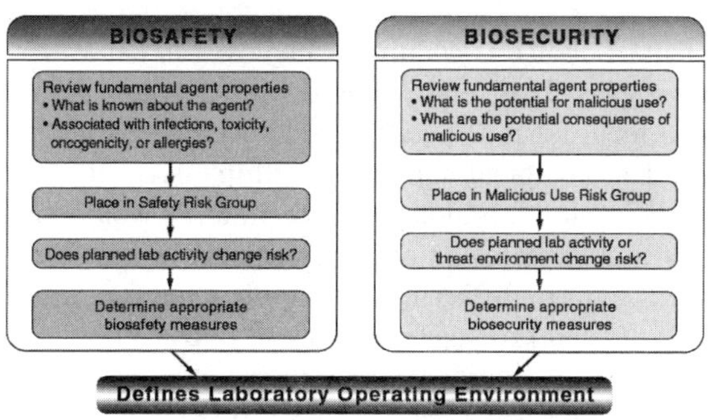

FIGURE 2.2 (See color insert following page 16.) Parallels between biosafety and biosecurity risk assessment methodologies.

There are multiple ways to acquire most biological agents. Most agents may be isolated from natural reservoirs and disease outbreaks in the environment. They may be created by *de novo* synthesis, turning an avirulent strain into a virulent strain, or other synthetic biology approaches. Or, they may be stolen from a laboratory, culture collection, or other legitimate facility.

The ability to successfully isolate biological agents from the environment depends on the adversary knowing where to collect the agent, how to identify the source, and how to select a virulent strain, as well as the availability of appropriate laboratory protocols within the open literature. For instance, an adversary is likely to encounter a number of technical obstacles in isolating a dangerous strain of *Yersinia pestis*, the bacterium that causes plague, from the environment. Despite the presence of large numbers of endemic plague foci on most continents, together the foci cover only about 6–7% of the Earth's land mass and cannot be considered widespread.[9] Scattered throughout these foci are many different strains exhibiting tremendous amounts of genotypic and phenotypic diversity. In addition to genetic variation in the bacterial genome, there is also extensive variation in plasmid content between strains. Consequently, the strains vary widely in virulence potential, and it is still not feasible to know what combination of genotypic and phenotypic traits are indicative of virulence in humans. As a result, successful isolation of BW-suitable strains from nature, in a non-outbreak situation, would be extremely challenging. An indication that extremely hazardous strains may not be very common in nature is that, in 2003, only 182 deaths attributed to plague were reported worldwide, even though many of the foci overlap with sizable human populations.

Theft of the biological agent from a legitimate facility, which often contains well-characterized agents, eliminates many of the difficulties associated with isolating a virulent strain from nature. However, the act of stealing the biological agent from a legitimate facility has a separate set of task-complexity factors to consider. The complexity of theft is based on both the characteristics of the facility's site and the facility's security posture (including physical security and security procedures). The ultimate objective of this risk assessment methodology is to aid a facility in analyzing its risk of being targeted for theft of its biological agents. Thus, it is important to analyze other plausible acquisition pathways to determine how attractive the facility is as a source of that particular agent.

Once the agent of choice is acquired, the adversary must be able to develop the agent into a weapon. This requires producing a suitable amount of the agent in the appropriate form for it to be effectively used as a weapon. Producing the desired quantity of a biological agent is a multiple-step process. Following the acquisition of a seed culture, a bioterrorist technician would have to inoculate a flask or fermenter that contains an appropriate growth medium. Different types of biological agents require different media in which to multiply. Although most bacteriological agents can propagate outside of a host, viral agents cannot; consequently, viruses require additional, sometimes highly sophisticated, treatments to stimulate growth. In general, the necessary growth media, supplies, and equipment are all easy to acquire, and many of these items may be obtained in prepackaged kits. The degree of difficulty associated with this process influences both agent selection and the dissemination scenario, which ultimately determine the potential consequences.

For instance, there is a great deal of variation in the required technical skills associated with amplifying an agent, producing quantities of liquid agent, and producing freeze-dried agent. Many agents lose virulence during amplification. This may be due to genetic mutations or even amplifying contaminants instead of the desired agent. It appears that the desired BW characteristics of many agents are highly susceptible to very small deviations from a specific set of growth condition, minus mutation or contamination. Thus, it is critical that the technician be able to conduct basic quality-control testing and evaluation. Because agents are susceptible to environmental degradation, special precautions need to be taken to preserve the integrity of the agents during amplification and storage. Liquid formulations may require stabilizers, nutrients, or other additives, whereas dry formulations usually are a lyophilized preparation. If a BW preparation cannot be stored until the optimal time of use without losing its efficacy, it may be less appealing to the adversary. Moreover, covert BW activities require adequate containment and safety measures to prevent accidental release or exposure of the workers; a biosafety accident could result in the detection of the program. To alleviate this operational security issue, adversaries may choose a production location in an area where the disease is endemic.

The chosen mode of dissemination of a BW depends on the effective routes of exposure for a particular pathogen or toxin. Pathogens and toxins may pass through the skin (typically through an abrasion), be inhaled or ingested orally, or transmitted by an insect vector. The mode of dissemination is an important factor in the complexity of the adversary's bioterror operation. An adversary must also consider the stability of an agent during and after dissemination. Stability can be affected by humidity, pH, temperature, chlorination, and ultraviolet radiation, among other factors.

The agent task complexity must be reevaluated at regular intervals. Advances in technology lower the agent task complexity associated with acquiring, producing, and effectively disseminating biological agents. Research publications, patents, and Internet-based surveillance and reporting of disease outbreaks provide increasing amounts of information to identify specific facilities or regions where the virulent agent of choice may be obtained. Over time, more individuals will acquire the skills and technologies to create dangerous pathogens through chemical synthesis and genetic engineering. The cost to acquire these skills and technologies will also gradually decrease over time. Some viruses have been synthesized in legitimate laboratories, and pathogenic bacterial agents could be engineered through the addition of virulence plasmids to a vaccine strain. Most recently, the 1918 pandemic influenza virus has been recreated in the laboratory.[10] The open exchange of scientific information also makes it easier for an adversary to learn production and dissemination protocols; instructions for amplifying and processing biological agents can easily be found on the Internet.

The fundamental properties of biological agents are only a starting point to assess the risk posed by the biological materials at a facility. From a security perspective, it is important to consider whether there are other factors that lower the task complexity of producing and disseminating the agent for potential adversaries.

If the facility has significant quantities of material, the adversary's task complexity associated with producing a suitable quantity of material will likely be lowered. Quantity may be especially important for materials that are nonreplicating, including toxins and special preparations. This would include a lyophilized agent preparation that has been treated to enhance its ability to form small aerosols. The risk assessment should incorporate the form of the material (e.g., aerosol preparation), the manner of storage (e.g., whether the material is prepared for long-term storage), and the quantities of nonreproducing materials (e.g., toxins) that are kept on hand.

In other words, consideration should be given to those factors that may lower the threshold of development for someone intent on malicious use. If an adversary can acquire a suitable quantity of material in a suitable form, the potential lowering of the task complexity for production and dissemination must be balanced with a potential increase in task complexity for acquisition. In most scenarios, it will be more difficult for an adversary to acquire a larger quantity of material because of the increased difficulty of removing large volumes of material covertly. After consideration of these additional agent task-complexity factors, the evaluation of acquisition, development, and dissemination should be revised accordingly.

2.2.1.2 Potential Consequences of Malicious Use

For laboratory biosecurity, we are concerned about the theft and use of dangerous biological agents in bioterrorism. Certain biological agents can cause highly infectious disease, which can jeopardize national and international security and stability. Thus, the consequences for biological agents are measured on a scale that captures potential national and international repercussions. Specifically, the consequences of using a biological agent maliciously are a function of four criteria:

1. Population impact: an estimate of the number of people who would be killed or injured by the malicious use of the biological agent.
2. Economic impact: an estimate of the economic loss that would be directly associated with the malicious use of the biological agent.
3. Psychological impact: an estimate of the degree to which public behavior or the public's perception of risk would be affected by the malicious use of the biological agent.
4. Operational impact: an estimate of the effect that the theft and misuse of the biological agent would have on the operations of the facility.

Inherently, dependencies exist among these consequence parameters. A higher population impact will necessarily lead to higher economic impacts. If people isolated themselves to avoid a contagious disease and did not report to work, the economic consequences would increase beyond those directly related to the bioterror attack itself.

The potential population impact is influenced by how easily the agent is transmitted (contagiousness), the percentage of those exposed who contract the disease and the resulting severity of disease (morbidity), and the percentage of those exposed who will die (mortality). Morbidity and mortality rates will be dose-dependent, whereas the dose will be scenario-dependent; the dose-dependent rates are often unknown. For all of these reasons, it may be most appropriate to use the morbidity and mortality rates associated with the natural progression of the disease. Although the natural disease parameters may not be adequate for determining the true impact of a bioterrorism attack, they provide comparable parameters to analyze the relative risks of the biological agents.

There will be direct economic impacts associated with any biological attack in the form of clean-up costs. The U.S. Environmental Protection Agency (EPA) spent over $42 million to clean up government buildings in Washington, D.C. following the anthrax attacks of 2001. The U.S. Federal Bureau of Investigation (FBI) estimated that the total damages of those attacks exceeded $1 billion.[11] Direct economic impacts will also be caused by agricultural bioterrorism that results in the loss of livestock or crops. There is likely to be a variety of less easily estimated secondary impacts, such as travel restrictions, market fluctuations, agricultural trade sanctions, extended closures of businesses, or long-term medical care.

A biological attack on humans or agriculture could result in varying degrees of anxiety, panic, and social disruption. How much a biological attack affects behavior and psychological well-being may be an important factor in the choice of biological material for certain adversaries. Biological agents that are only suitable for an isolated attack on a specific target, such as an individual, will likely not cause public panic. For example, when Bulgarian dissident Georgi Markov was assassinated by injection with a ricin pellet, it was viewed as an attack on that specific individual, not an event that might threaten the larger population. In contrast, the 2001 anthrax letter events in the U.S. directly impacted only a few isolated populations on the East Coast, yet there were psychological consequences throughout the U.S. and the world. The impact led to changes in behavior by large segments of the general East Coast population because they feared a repeat attack. And countries around the world experienced a significant increase in the number of "white-powder" incidents.

Following a bioterrorism event, public anxiety may be directly related to the availability of medical countermeasures. A contagious disease, such as smallpox or pneumonic plague, would likely have more psychological impact than a disease that cannot be transmitted from person to person. A disease for which medical countermeasures do not exist, such as Ebola or Marburg, would likely cause considerably more public anxiety than a treatable infectious disease. Further, it is likely that biological agents that cause human disease will have a significantly greater psychological impact than biological agents that cause only animal or plant disease.

Theft and use of a biological agent is unlikely to have operational impacts on a specific facility because the biological agent will generally not be attributable to that facility. If illicit acquisition from a legitimate facility is discovered or suspected, all facilities may be operationally impacted by additional regulatory controls. The scope of these potential impacts is unknown.

For biological agents that have been modified through experimentation, the risk assessors should consider evaluating those materials as separate agents with different characteristics. Examples of these procedures include enhancing antibiotic resistance, creating modifications that circumvent existing vaccines, and increasing environmental stability. Agents that have been genetically modified in a way that affects virulence, stability, or other weaponization characteristics — particularly if it results in materials that, if used maliciously, would cause more significant consequences than similar material isolated from nature — should be evaluated separately.

An analysis of the agent task complexity and potential consequences can be used to place biological agents in malicious use risk groups (Table 2.1). We define these agent risk groups as Low Malicious Use Risk (LMUR), Moderate Malicious Use Risk (MMUR), High Malicious Use Risk (HMUR), and Extreme Malicious Use Risk (EMUR).

2.2.2 Evaluate the Potential Adversaries

The probability that an adversary will target a bioscience facility to steal and then subsequently misuse a biological agent cannot be known. Historical data, even when available, cannot reliably predict whether an adversary will attack. What happened in the past may not necessarily happen in the future. As a result of this limitation, we evaluate instead an adversary's attributes that enable him or her to pose a threat to a biological agent. Specifically, we measure the attributes of each potential adversary by considering a function of three criteria:

1. Motive — Why would the adversary try to steal a biological agent and use it maliciously?
2. Means — Does the adversary have the ability to successfully steal a biological agent and use it maliciously?
3. Opportunity — Can the adversary steal the biological agent covertly or must he steal it overtly?

Laboratories generally should not attempt to evaluate these "adversary attributes" without consulting external experts.[12,13] Traditionally, security and law enforcement (LE) services maintain information on specific adversaries and their attributes. Thus, laboratories should develop collaborations with such agencies. Appendix A has a set of questions that can be used to facilitate interactions with LE to help characterize adversaries who may threaten biological agents.

Defining notional adversaries that span the spectrum of plausible adversaries is one mechanism to facilitate analysis of their attributes. Adversaries can

TABLE 2.1
Malicious Use Risk Groups for Biological Agents

Malicious Use Risk Group	Task Complexity	Potential Consequences	Examples
Nonpathogenic	Not applicable	Little or no consequences if used maliciously	Noninfectious forms of pathogens (e.g., inactivated organisms and nucleic acids), nonpathogenic strains, extreme halophiles, and extreme thermophiles
Low	High — may be difficult to acquire, are hard to produce in sufficient quantities, and unsuitable for dissemination	Low — low population impact, inflict little economic damage, and are expected to have a low psychological impact on the general population	*Mycobacterium leprae,* small quantities of toxins (i.e., less than the threshold for the Select Agent list), agents transmitted primarily by parenteral or sexual exposure (e.g., malaria, hepatitis, and gonorrhea), attenuated strains, genetic host strains of *Escherichia coli*, measles, mumps, and *Pseudomonas aeruginosa*
Moderate	Moderate — many of the steps associated with successfully deploying these agents will be relatively easy, but perhaps one or two of the critical steps are difficult	Moderate — localized consequences with low-to-moderate casualties, moderate-to-significant economic damage, and the potential to cause pervasive anxiety	*Coccidioides immitis,* agents that pose a threat primarily through food (e.g., salmonella, *E. coli O157:H7*, and shigella) or water (e.g., *Vibrio cholerae* and *Cryptosporidium*), and larger quantities of some toxins
High	Low — not particularly difficult to deploy as weapons	Moderate to high — national or international consequences, moderate-to-high casualties and/or economic damage, and the potential to cause mass panic and significant social disruption	*Bacillus anthracis, Francisella tularensis, Coxiella burnetti,* foot-and-mouth disease virus, and *Yersinia pestis*
Extreme	Same as high but they receive a higher classification because they are not found in nature	Same as high	Variola major virus, and could include genetically engineered agents if they were suspected of representing a high risk

be defined on the basis of their motives: terrorists, extremists, or criminals. Adversaries can also be defined on the basis of their access to the facility: those that have legitimate access to the facility (insiders) and those who do not have authorized access (outsiders). If a facility has access controls in place, insiders can be further subdivided based on access, e.g., insiders with full access, insiders with building access, and insiders with site access. Specific examples of notional adversary descriptions that may helpful in conducting a risk assessment can be found in Appendix B.

In addition to simplifying the analysis, notional adversaries may be the only option in the absence of intelligence that indicates there are specific adversaries who are targeting a facility. If a facility can identify a specific adversary who is unique to that facility, then that specific adversary can and should be included in the risk assessment. In order to examine how much of a threat each adversary poses, various elements of the adversary's motive, means, and opportunity should be examined.

2.2.2.1 Motive

A report published by the Monterey Institute of International Studies describes the results of a comprehensive review of 33 alleged incidents involving biological agents from 1960 to 1999.[14] The review articulated a number of reasons why terrorists and criminals have sought to use BWs. In descending order of importance, the authors found that the most important motives for potential bioterrorists were

1. Revenge
2. Nationalist/extremist objectives
3. Apocalyptic ideology
4. Government protest
5. Animal activism
6. Miscellaneous (antiabortion, environmental terrorism, political dominance, industrial sabotage, and assassination/murder)

As this list shows, different adversaries may want to acquire a biological agent for different reasons. A criminal may be interested in economic gain or revenge. An extremist may want to make a political statement. A terrorist may want to pursue mass murder or mass hysteria. The particular motivations of an adversary may lead him or her to target particular agents that would be most beneficial for his or her objectives. For instance, the motives of a terrorist would not likely be met by stealing an low malicious use risk agent, and a criminal or extremist may not need to steal an high malicious use risk agent to achieve his objective.

The level of an adversary's motivation to acquire a biological agent from a particular facility is dependent on how attractive the particular facility is as a source of that agent. Does the same agent exist in other facilities or the environment?

Does the adversary have the option of stealing the agent from other facilities, or successfully isolating it from nature? The nature of the facility will also influence the adversary's motive. For instance, a terrorist may be more motivated to attempt to steal a high malicious use risk agent from a research facility than a diagnostic facility where HMUR agents may be occasionally identified but are not stored.

Despite the fact that historical data cannot be used to predict the future, it does provide some context and should be used judiciously in evaluating the threat that such an event will occur again. Therefore, it is important for risk assessors to establish whether this type of attack has occurred in the past and/or whether there is a history of activity in the area by a particular adversary.

2.2.2.2 Means

The threat that an adversary presents to a facility is also influenced by whether the adversary has the means to steal a biological agent and use it maliciously. The adversary needs adequate technical skills, operational knowledge, and tools to conduct an act of bioterrorism. Some adversaries will be better suited than others to successfully execute such an attack, and they should be ranked accordingly.

Even if the adversary is highly motivated to act, he or she needs the means (training and tools) to execute the scenario and achieve the desired consequences. If the adversary lacks adequate technical skills, operational knowledge, and tools, he or she will only pose a limited threat to the facility. In order to evaluate the means of an adversary to steal and misuse a biological agent, a certain set of capabilities must be assumed or defined. In general, these capabilities are established by working with local law enforcement and intelligence officials. These capabilities should include the level of scientific knowledge and access to scientific equipment and technology, knowledge of facility operations, knowledge of where the biological agent is located, knowledge of how it is protected, and necessary tools to overcome the existing protection measures.

Different adversaries may have different means to steal and use a biological agent maliciously. For instance, an Insider with Building Access may have knowledge of facility operations, scientific knowledge, and knowledge of where a particular agent is stored, but he or she may not know how it is protected. A Terrorist Outsider may have extensive scientific knowledge and access to scientific equipment, and the ability to gain access to the facility forcefully, but he or she may not know where the agent is located or how it is protected.

2.2.2.3 Opportunity

An adversary's opportunity to steal and misuse a biological agent can be assessed by evaluating the adversary's level of access to the agent. In particular, opportunity is determined by an evaluation of the extent to which the adversary can steal the biological agent covertly. The opportunity attribute of the adversary is low if he or she must rely on stealing the agent overtly, which by definition makes it more likely that the adversary will be detected than if he or she could steal the agent covertly.

For instance, an Insider with approved access to a laboratory that works with a biological agent will have more opportunity to steal that agent covertly than an Insider who has access to the building but not that particular laboratory. However, that Insider with Building Access may have more opportunity to steal that agent covertly than a Terrorist Outsider who does not have legitimate access to the facility's site or its buildings.

A Terrorist Outsider may choose to collude with an Insider with Full Access to steal the biological agent; collusion may be voluntary or involuntary. Collusion increases not only the Terrorist Outsider's opportunity to steal the biological agent but also increases the risk of detection for the Terrorist Outsider. In general, the more individuals who are involved in an illegitimate act, the more likely it is that their act will be discovered.

2.3 EVALUATE SCENARIOS

Now that the risk assessment process has evaluated the biological agents at a facility and defined the notional adversaries who may want to steal and misuse those biological agents, it is necessary to create and evaluate various biosecurity scenarios based on those agents and adversaries.

2.3.1 Create Scenarios

Earlier in this chapter, we indicated that the biosecurity risk assessment process will concentrate on evaluating the threat potential and consequences of specific scenarios, which include three components:

1. A specific pathogen or toxin
2. An individual or group of individuals who wish to steal a pathogen or toxin from a bioscience laboratory
3. The theft of a pathogen or toxin

In theory, scenarios should be created by pairing every possible adversary with every biological agent. A more detailed discussion of creating scenarios is described in Appendix B. An analysis of this magnitude may be overwhelming for many bioscience facilities because of the number of biological agents and possible adversaries. In these cases, a screening process can be used to reduce the scope of the risk assessment.

The purpose of this screening process is to limit the number of biological agents that continue through the risk assessment process to only those that may be of interest to an adversary. If an agent does not meet a particular threshold — such as a moderate malicious user risk agent — it can be "screened out" of the risk assessment, and protection priorities can be focused elsewhere. The rationale for screening out a particular biological agent should be documented in the risk assessment.

After the agent screening process, notional adversaries should be paired with the remaining biological agents, assuming that the notional adversaries have been defined as having the necessary attributes to successfully steal and misuse those agents. For example, if the facility has high malicious use risk materials, three scenarios may need to be constructed — one each to evaluate theft and use by an Insider, a Terrorist Group, and a Single Terrorist. The defined attributes for the criminals and the political extremists may not be adequate or appropriate for those adversaries to steal and misuse high-risk agents; these scenarios may be screened from the full assessment. Terrorists should be included in scenarios associated with high and extreme risk biological agents. Terrorists should also be included in scenarios with moderate malicious use risk materials if the facility environment indicates a high level of activity in the area, if this facility has the most dangerous biological materials in the nation and the terrorists have a national presence, and especially if locally active terrorists have expressed any interest in biological agents. Insiders should be considered in every biological agent theft-and-use scenario.

2.3.2 Assess Vulnerabilities

Following the development of biosecurity scenarios that the risk assessment will specifically evaluate, the site or facility should be evaluated for vulnerabilities. There are a variety of meanings associated with the phrase *vulnerability assessment.* The concept of vulnerability assessment used in this book is dependent on the defined scenarios. In other words, risk assessors should consider what vulnerabilities exist at the facility that would allow the defined scenarios to occur.

There are many software modeling tools available for a vulnerability assessment, most of which focus on very-high-consequence assets (e.g., special nuclear materials or weapons), the outsider adversary, and the ability of the security system to detect the intrusion and to delay the intruder long enough for the response force to intercept the adversary. These tools presume a level of adversary capability and security-system sophistication that generally exceeds what would be applicable to bioscience facilities.

For the biosecurity risk assessment, the efficacy of the facility's laboratory biosecurity system should be evaluated for vulnerabilities. In effect, this step in the risk assessment provides site-specific information that can counter the adversaries' attributes that were defined in Section 2.2.2. A facility with many biosecurity vulnerabilities improves the adversary's chance of success, whereas a facility with few biosecurity vulnerabilities minimizes the adversary's ability to successfully steal a biological agent.

Specifically, site biosecurity vulnerability is equivalent to a function of the vulnerabilities in the following components of a laboratory biosecurity system:

1. Physical security
2. Personnel security
3. MC&A
4. Transport security
5. Information security
6. Program management

Chapter 3 ("Components of Biosecurity") and Chapter 4 ("Program Management") will describe these fundamental components of a laboratory biosecurity system in greater detail; this section illustrates how each component impacts the vulnerability from Outsider and Insider Adversaries. Appendix A includes examples of questionnaires that can be used to help conduct site vulnerability assessments.

For physical security, it is important to consider whether the site controls individuals' access to those buildings and laboratories where the specific biological agents in the scenario are stored and used. In those scenarios that involve an adversary who does not have legitimate access to those areas where the biological agent is stored or used, risk assessors should also determine whether there are systems in place to detect unauthorized intrusion. The existence and efficacy of perimeter fences and response forces should also be evaluated for those scenarios that include an Outsider who does not have legitimate access to the site or a particular building.

Personnel security measures, such as background screening of individuals for reliability and trustworthiness prior to granting access to specific buildings and laboratories, represent a critical biosecurity measure. In fact, personnel security is the most important way in which a bioscience facility protects against an Insider with Full Access stealing a biological agent. The efficacy of personnel security measures is not a relevant protection measure against an Outsider adversary.

If a facility has a reliable MC&A system in place, it is more difficult for an Insider to covertly steal a biological agent. Transport security measures, which are aimed at maintaining protection of biological agents while they are moved from one secured laboratory to another, reduce the opportunity of both the Insider and Outsider adversary to steal a biological agent. The protection of information about the specific elements of the biosecurity system (e.g., where certain biological agents are stored and used, how they are protected) increases the difficulty for Outsiders and Insiders without Full Access. Biosecurity program management reduces a facility's vulnerabilities against all relevant adversaries.

A reliable quantitative assessment of the effectiveness of a biosecurity system is not possible. For example, we cannot quantify how often an Insider abandons his or her attempted theft of a biological agent because he or she believes he or she will be detected. Quantifying the value of the security system is also complicated because detecting the covert theft of biological materials by an the Insider is extremely difficult. Although there are many biosecurity system recommendations made in this book that are designed to deter the adversary from stealing a biological agent, effective deterrence (e.g., the absence of malicious actions) is not measurable. Therefore, the vulnerability assessment portion of the risk assessment described in this model is based upon establishing the difference between the recommended level of biosecurity and the assessed level of biosecurity at the site.

NOTE: Vulnerability assessments generally should be protected as sensitive information and, in the U.S., may be protected from release under the Freedom of Information Act (FOIA) using the Category 2 Exemption.[15]

2.4 CHARACTERIZE THE RISK

Characterizing the laboratory biosecurity risk consists of three specific steps:

1. Evaluate the threat potential and consequences of each scenario.
2. Determine which scenarios represent acceptable risks and which represent unacceptable risks.
3. Develop a risk statement or definition to articulate the objectives of the biosecurity system.

2.4.1 Evaluate Threat Potential and Consequences

At the beginning of this chapter, we indicated that biosecurity risk is a product of threat potential and consequences of specific biosecurity scenarios. Because we have limited our biosecurity objective to protecting biological agents from theft, consequences are also limited to the consequences associated with the malicious use of those biological agents. Section 2.2.1.2 specifically indicated that consequences are equivalent to a function of the population, and the economic, psychological, and operational impacts of using a specific biological agent maliciously.

What remains is to define the *threat potential* part of the biosecurity risk equation. Specifically, the threat potential of a specific scenario is a function of three elements of the scenario that were each explained earlier in this chapter:

1. Agent task complexity (Section 2.2.1.1)
2. Adversary attributes (Section 2.2.2)
3. Site vulnerability (Section 2.3.2)

With values for threat potential and consequences, each scenario can then be plotted on a two-dimensional biosecurity risk graph with hyperbolic curves that distinguish between levels of relative risk. Figure 2.3 shows the three elements of threat potential and how they contribute to biosecurity risk.

Figure 2.4 depicts the entire biosecurity risk equation, with its individual elements, as it has been explained in this chapter.

Appendix B provides additional details on the biosecurity risk-assessment methodology described here.

2.4.2 Define Acceptable and Unacceptable Risks

The final step of the formal biosecurity risk assessment is the plotting all of the scenarios — each with a particular value established for threat potential and consequences — on a two-dimensional biosecurity risk graph. The analytical work of assessing risk has been completed. It is important for management to appreciate that the scenarios reflect relative biosecurity risks at that facility; the scenarios are not definitive determinations of risk and cannot be compared to biosecurity risks at other bioscience facilities or to assessed risks in other industries.

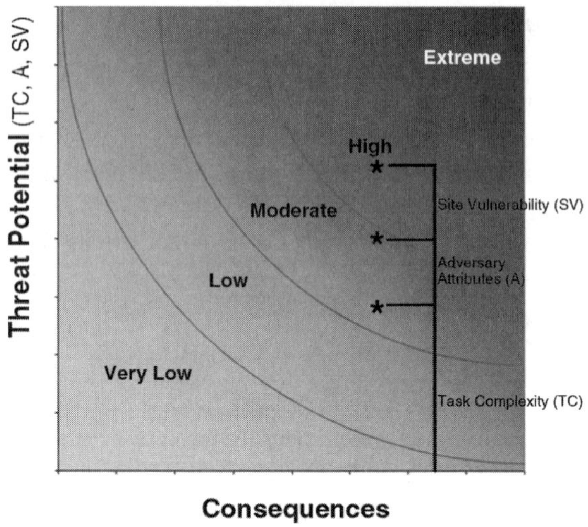

FIGURE 2.3 **(See color insert following page 16.)** Elements of biosecurity risk.

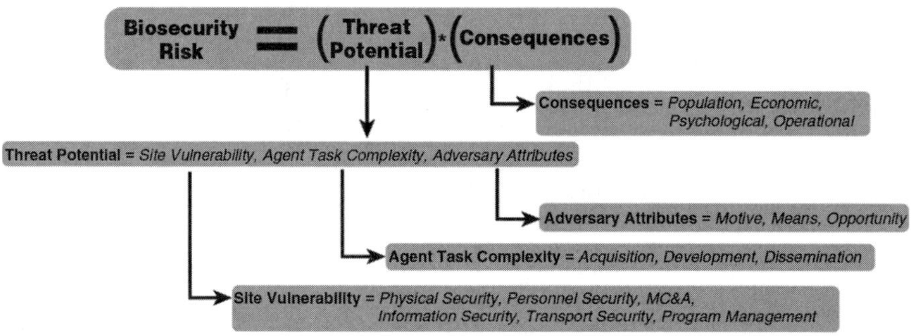

FIGURE 2.4 Biosecurity risk flowchart.

Now management must compare these scenarios against each other and determine which biosecurity risks are acceptable and which are unacceptable. The identified "unacceptable risk" scenarios must be protected against by mitigating the vulnerabilities of the biosecurity system that would allow those unacceptable scenarios to occur. The identified "acceptable risk" scenarios may require little or no additional biosecurity protection.

Resources for laboratory biosecurity will always be limited, and existing resources should be used as efficiently as possible. It stands to reason that higher-risk scenarios should receive more protection than lower-risk scenarios, and that the cost of implementing biosecurity should be proportional to the risk. In other words, some biosecurity scenarios may be more unacceptable to management than others, and some biosecurity scenarios may be more acceptable than others. Instead of simply making a binary decision about unacceptable and acceptable risk scenarios, management may prefer to prioritize the biosecurity scenarios that emerge from the risk assessment.

This risk decision will reflect management's level of risk tolerance as well as the availability of its resources. If management is risk-tolerant, it will consider the majority of the biosecurity scenarios identified in the risk assessment as acceptable risks and thus will not need to invest substantial resources in biosecurity systems. Conversely, if management is risk-averse, it will judge most of the biosecurity scenarios to represent unacceptable risks. However, in this case, management will be required to invest substantially in biosecurity systems. Therefore, this decision about which biosecurity risks are unacceptable represents a critical step in determining how resources will be allocated to mitigate them.

Biosecurity scenarios involving Insiders tend to pose a higher risk than scenarios involving Outsiders. Insiders have access to the facility and the buildings and laboratories where the biological agents are used and stored. Moreover, Insiders have knowledge of the facility operations as well as its security system. Some Insiders also have relevant technical skills and know precisely how to covertly remove a specific viable and virulent organism from the facility. Biosecurity scenarios involving Outsiders tend to pose a lower risk than Insider scenarios because of the following facts: most pathogens can be isolated from nature; most Outsiders will not know exactly where the viable and virulent agents are stored; most Outsiders have limited knowledge of the facility's operations and security system; and in general Outsiders would have to rely on colluding with an Insider to covertly steal a pathogen, which increases the Outsiders' risk of detection. Moreover, most bioscience facilities do not have biological agents that are unique or valuable enough as a weapon to warrant the risk an adversary would incur by stealing and trying to disseminate those agents maliciously.

Management should also appreciate that the cost of protecting against a biosecurity scenario involving an Insider is generally lower than the cost of protecting against a biosecurity scenario involving an Outsider overtly attacking the facility. After comparing the low risk against the high cost of protecting against such an attack, many facilities may decide to accept that risk. Others may decide that the risk is high enough to warrant addressing it through incident response planning. Still others may decide to protect themselves against such an attack either because the agents they possess warrant such protection or because their management believes, for any number of reasons, that the facility must implement an expensive, risk-averse security posture.

Whatever position management takes regarding acceptable and unacceptable biosecurity risks should be documented to articulate the facility's specific biosecurity protection objectives. This "risk statement" should clearly define which biosecurity scenarios represent unacceptable risks that the biosecurity system must comprehensively mitigate. This statement may specify that the risk of some of the biosecurity scenarios could be reduced simply through incident response planning. This statement may also indicate that certain biosecurity scenarios represent acceptable risks that require no biosecurity protection.

The risk statement should also indicate what specific protection strategy the laboratory biosecurity system should employ to mitigate the unacceptable risk scenarios. In general, there are three different security strategies for management to consider:

1. Deny: prevent the adversary from gaining access to a particular biological agent.
2. Contain: prevent the adversary from leaving the facility while in possession of a stolen biological agent.
3. Deter: discourage an adversary from stealing a biological agent by making theft of that agent appear very difficult or impossible.

Denial may be achieved by implementing security measures of sufficient strength to defeat the capabilities of the adversary posing the risk, or by detecting the attack, responding, and intercepting the adversary prior to the adversary reaching the biological agent he or she intends to steal. Containment requires the detection of the attack and a response capability sufficient to prevent the adversary from leaving the facility's jurisdiction.

Denial is an appropriate strategy for protecting against those Outsiders who do not possess adequate tools or knowledge to overcome the facility's security measures; however, the more sophisticated the adversary, the more costly denial becomes. A containment strategy against Outsiders in theory would be less expensive than a denial strategy, but it also would be quite resource-intensive and is only recommended when the risk assessment indicates that an outside adversary poses a very high risk.

Because the Insider generally poses a higher threat to biological agents at a bioscience facility than Outsiders, deterrence is usually the most appropriate protection strategy. In fact, because of the difficulty of detecting theft of biological agents by Insiders, deterrence is often the only option for protecting against the Insider. Denial may be used, at a reasonable cost, to counteract those Insiders who have no need for legitimate access.

As the objective moves from denial to containment to deterrence, the level of risk the facility must accept increases. However, there is little justification for incurring the expense associated with denying an adversary the ability to acquire a given pathogen from a particular facility when that pathogen is available in nature.

The risk statement, which articulates the biosecurity system's objectives and protection strategy, should precede the design and implementation of the biosecurity system. It should also be used as the basis against which the effectiveness of the existing system is evaluated for vulnerabilities and upgrades. As such, it will also become the basis for allocating the facility's biosecurity resources.

2.5 RISK REDUCTION

Risk reduction is the process of considering, designing, and implementing various means to mitigate the risks identified in the biosecurity risk assessment. Biosecurity risk reduction can be summarized by the following steps:

1. Identify options for changing the existing protection measures to prevent the unacceptable risk scenarios from occurring.
2. Design and implement or upgrade the system to specifically address the unacceptable risk scenarios and the articulated protection strategy.
3. Develop a written security plan to document the objectives and function of the biosecurity system.
4. Develop written incident response plans to address both the acceptable and unacceptable risk scenarios.
5. Provide training on both the security and incident response plans.

Steps one and two are covered in Chapter 3, and steps three through five are discussed in Chapter 4.

Risk can be reduced in two ways: lowering the threat potential of a particular scenario that implements security measures, and/or lowering the consequences of that scenario. The mission of most biodefense programs is to reduce the consequences of the next bioterrorism event by strengthening medical countermeasures (vaccines and therapies); improving surveillance, detection, and diagnostic methodologies; and enhancing public and agricultural health infrastructure. This book focuses on reducing the threat potential of these scenarios by implementing laboratory biosecurity systems.

As described in Section 2.4.1, threat potential is a function of agent task complexity, adversary attributes, and site vulnerability. Implementing a laboratory biosecurity system can reduce the site's vulnerabilities and alter the adversaries' motives, means, and opportunity. However, implementing a laboratory biosecurity system cannot affect the agent task complexity component of threat potential or the consequences of using a particular biological agent maliciously. Thus, once resources are expended to address the risks the facility deems to be unacceptable, there will still be residual risk — the unmitigated risk — that the facility must accept. The facility must understand that there will always be residual risk regardless of the amount of money spent on biosecurity measures. There is no way to design the perfect security system that would reduce the level of risk to zero. Figure 2.5 shows the difference between the absolute risk of a particular scenario in the absence of any medical countermeasures or laboratory biosecurity and the residual risk after implementing biosecurity and accounting for medical countermeasures.

In theory, biosecurity systems should be designed to address specific scenarios that management has determined represent unacceptable risks. The laboratory biosecurity measures to reduce the risk of an Insider scenario may be very different from the measures to reduce the risk of an Outsider scenario. However, it is important for managers and designers of biosecurity systems to appreciate that risk-reduction measures often can reduce the risk of more than one scenario. Implementing access-control and intrusion detection systems as well as MC&A and transport security to deter the Insider threat will also provide a degree of effectiveness against the Outsider threat. Figure 2.6 demonstrates that implementing biosecurity measures designed to counter the Insider threat potential

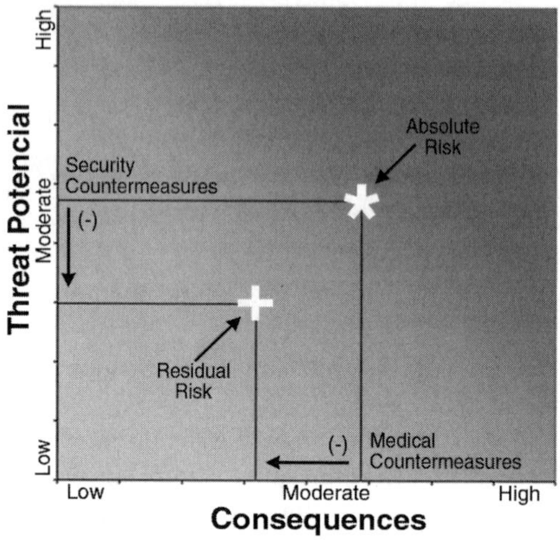

FIGURE 2.5 **(See color insert following page 16.)** Risk reduction.

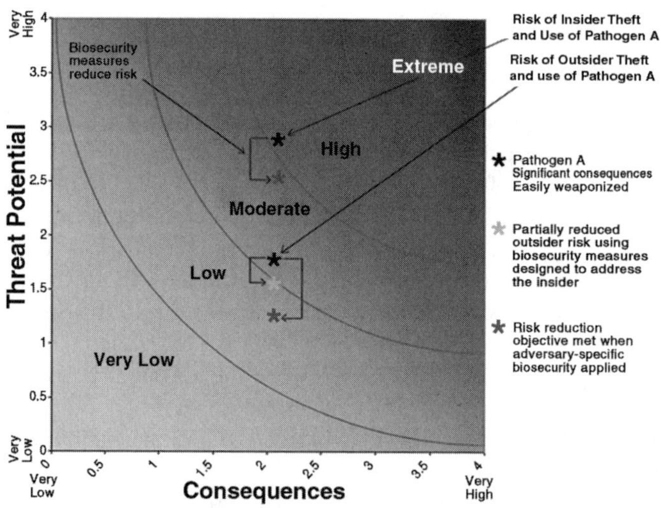

FIGURE 2.6 **(See color insert following page 16.)** Insider vs. outsider threat reduction.

partially reduces the Outsider threat potential as well. In the case of biosecurity, the inverse may not necessarily be true. Biosecurity measures specifically implemented to reduce the Outsider threat potential may not substantively reduce the Insider threat potential.

REFERENCES

1. Kaplan, S. and Garrick, B.J., 1981, On the quantitative definition of risk, *Risk Analysis*, 1(1).
2. U.S. Environmental Protection Agency, Integrated Risk Information System, http://www.epa.gov/iris/intro.htm.
3. Stern, P.C. and Fineberg, H.V., Eds., 1996, National Research Council, Committee on Risk Characterization, *Understanding Risk: Informing Decisions in a Democratic Society,* Washington, D.C.: National Academy Press, available online at: http://darwin.nap.edu/books/030905396X/html.
4. The U.S. Government Accountability Office has endorsed a risk management approach for addressing mitigating security threats, U.S. GAO, 2001, Homeland Security: A Risk Management Approach Can Guide Preparedness Efforts, Washington, D.C.: U.S. GAO-02-208T. Also see U.S. GAO, 2003, Combating Bioterrorism: Actions Needed to Improve Security at Plum Island Animal Disease Center, Washington D.C.: GAO-03-847.
5. U.S. FEMA, 2003, FEMA 426: *Reference Manual to Mitigate Potential Terrorist Attacks against Buildings*, Washington, D.C.: FEMA.
6. Belton, V. and Stewart, T.J., 2002, *Multiple Criteria Decision Analysis: An Integrated Approach,* Norwell, MA: Kluwer Academic.
7. World Health Organization, 2004, *Laboratory Biosafety Manual*, 3rd ed., Geneva: WHO.
8. Gaudioso, J., Salerno, R.M., and Barnett, N., 2006, Developing a risk assessment and risk management approach to laboratory biosecurity, *Applied Biosafety*, 24–31.
9. Anisimov, A.P. et al., 2004, Intraspecific diversity of *Yersinia pestis*, *Clinical Microbiology Reviews,* 17(2), 434; WHO data available at http://www.who.int/wer/2004/en/wer7933.pdf.
10. Salerno, R.M. and Hickok, L.T., forthcoming 2007, Strengthening Bioterrorism Prevention: Global Biological Materials Management, *Biosecurity and Bioterrorism,* 5(2).
11. Tumpey, T.M. et al. 2005, Characterization of the 1918 Spanish Influenza Pandemic Virus, *Science*, 310(5745), pp. 77–80.
12. Lengel, A., 2005, Little progress in FBI probe of anthrax attacks, *Washington Post*, September 16, p. A01.
13. Communication from the Commission to the Council and the European Parliament, 2003, On Cooperation in the European Union on Preparedness and Response to Biological and Chemical Agent Attacks (Health Security), Brussels, COM(2003) 320 final, http://europa.eu.int/comm/health/ph_threats/Bioterrorisme/com2003_320_en.pdf.
14. U.S. GAO, 1999, Combating Terrorism: Need for Comprehensive Threat and Risk Assessments of Chemical and Biological Attacks, Washington, D.C.: U.S. GAO, GAO/NSAID-99-163 (http//www.gao.gov).
15. Tucker, J.B., *Historical Trends Related to Bioterrorism: An Empirical Analysis.*
16. The Category 2 Exemption includes information that pertains solely to the internal rules and practices of the facility. This exemption category has two profiles: "high" and "low." The "high" profile permits withholding of a record that, if released, would allow circumvention of a facility rule, policy, or statute, thereby impeding the facility in the conduct of its mission (e.g., vulnerability assessments). The "low" profile permits withholding if there is no public interest in the record, and it would be an administrative burden to process the request.

3 Components of Biosecurity

Physical or engineered security is an important component of a biosecurity system and is often the first thing people think of when they hear the term "security." Sometimes that association has negative connotations, especially when those who understand biology are exposed to security systems that are dominated by traditional physical security that is not designed with any recognition of the unique attributes of biological materials. A comprehensive biosecurity program must include not only physical security but also personnel security, MC&A, transport security, information security, and program management elements (Figure 3.1). These components are the tools in the biosecurity officer's toolkit. The biosecurity officer can select the most appropriate implementation of each element based on his or her facility's unique risks. There are many resources available on the various components of a biosecurity system, especially physical security.[1]

Biosecurity measures should be more stringent for higher-risk scenarios than for lower-risk scenarios. Often, this means that higher-risk agents require more security than lower-risk agents. Low- and very-low-risk agents may require little or no security, whereas moderate-, high-, and extreme-risk agents should receive commensurately higher levels of security. Biosecurity measures should be implemented in a graded manner. A graded protection system is achieved by increasing security incrementally and forming concentric layers of protection around the facility's agents based on the results of the risk assessment. Each layer should have more physical security, personnel security, and MC&A requirements than the previous layer. Information and transportation security requirements will also vary based on the risk assessment, but these controls are not necessarily specific to physical areas within a facility.

This chapter describes each of the critical components of a laboratory biosecurity system. Chapter 4 discusses how to develop and maintain an appropriate biosecurity system using these elements, and Chapter 5 provides specific recommendations for creating biosecurity with combinations of these elements for facilities with different risks.

3.1 PHYSICAL SECURITY

3.1.1 Objective and Principles of Physical Security

Increasing physical security countermeasures is often the most obvious way to reduce the risk that an outside adversary poses and to minimize the threat of those insiders who do not require access to those assets the facility is protecting.

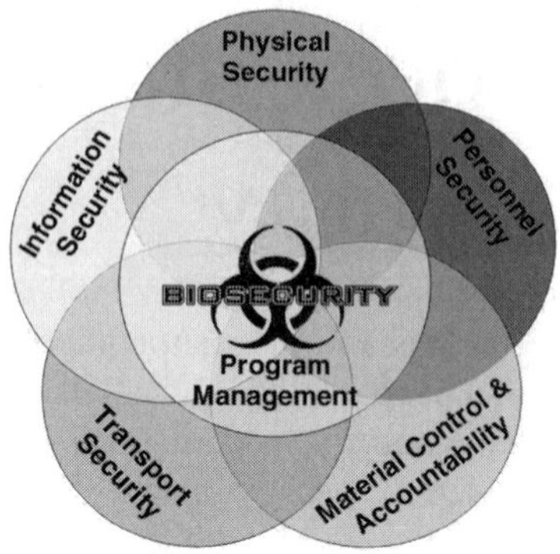

FIGURE 3.1 **(See color figure following page 16.)** Components of biosecurity.

Physical security seeks to reduce the risk of unauthorized access to specific areas or assets. This is accomplished with a mix of several fundamental elements: establishing boundaries, access controls, intrusion detection, and alarm assessment. Access control mechanisms include locks and other barriers to prevent unauthorized individuals from gaining access to restricted items or areas. Intrusion detection systems alert security personnel to attempts to gain access without authorization. Alarms must then be assessed to determine whether they are valid or are false alarms. Valid alarms that have been assessed as either an attempted or successful access to a restricted item or area by an unauthorized individual should be addressed by properly trained response personnel. Response is an important overall element of biosecurity and is addressed in Chapter 4.

3.1.2 Physical Security Elements

3.1.2.1 Perimeters and Other Boundaries

Boundaries must be established to demarcate the areas that are under some sort of access limitation. A fence defines the boundaries of the campus as well as providing a means to control personnel and vehicle access. For facilities that hold only low or moderate malicious use risk agents, signage may provide sufficient property demarcation. Boundaries to restricted areas can include walls, windows, doors, pass-through boxes, pass-through autoclaves, or other equipment access points. Entry and emergency exit doors should be mounted with the hinge pins on the inside of the restricted area, as should any hardware that is associated with securing the doors or windows, such as locks or handles. Exterior ladders should be secured to prevent unauthorized access to roofs and interior courtyards.

The overall layout of the facility and the locations of access control features are important to consider in order to ensure that the normal and emergency paths of employees and visitors do not inadvertently leave gaps in security boundaries. Paths should be analyzed to ensure that routes through applicable checkpoints are enforced without providing alternate, unsecured routes, and that emergency egress paths do not channel individuals into areas they would not normally have access to. The strength of the perimeter envelope protecting a restricted area will influence how long it takes an outside adversary to gain unauthorized access to the restricted area. The stronger the perimeter is, the longer the "delay" will be between the time of the initial intrusion alarm and the time it takes the outside adversary to gain access to the protected material. The longer the delay, the more opportunity the response force has to respond to an intrusion detection alarm.

3.1.2.2 Access Controls

Access controls provide reasonable assurance that only authorized personnel are allowed to enter a restricted area. The type of access controls selected depends on the level of surety required that only authorized personnel can enter a restricted area. Access can be controlled with a unique item, such as a physical or electronic key. To increase the reliability of the access controls, unique knowledge, such as positive identification by a guard or a personal identification number (PIN), can be required in addition to the unique item. The unique knowledge ensures that the individual who possesses the key is authorized to possess it.

This form of user authentication may also be achieved using biometric devices, which provide a higher level of assurance than a PIN. However, biometric devices are currently limited in utility because some individuals cannot "enroll," i.e., enter a biometric pattern that will subsequently be recognized as their own. This limitation means that any system incorporating biometric devices needs to maintain a separate system of authentication, such as guard identity verification, for those individuals who are unable to enroll in the biometric authentication system.

3.1.2.3 Intrusion Detection

Intrusion detection aims to notify facility staff that an unauthorized individual may have entered or tried to enter a restricted area. At its simplest, intrusion detection is an alert staff member who notices that something is amiss, such as a broken window or an open door that is normally closed. Facilities with only low malicious use risk agents can rely on training their staff to report these types of abnormal situations. Facilities with higher-risk agents can choose between roving guard patrols that are tasked specifically with monitoring the status of restricted-area entry points or electronic intrusion detection systems. Electronic systems have the advantage of providing constant monitoring. Mechanically based systems are inherently less effective than electronic systems but, depending on the facility's level of risk tolerance and local regulations, can be used when electronic systems are too costly or are not available.

Electronic intrusion detection systems are typically associated with electronic access control systems. If forced entry occurs, or if a door or other monitored entryway is open for an extended period of time, an alarm will be generated. The electronic network can be configured to detect tampering so that, if a communication line is cut or a junction box is tampered with, an alarm will be generated under these conditions as well. Glass-break sensors will send an alarm if a protected window is broken.

Other types of sensors, including motion detection sensors, will generate an alarm if they are triggered. Motion detection sensors, and any other sensor type not associated with detecting a breach in the boundary of the restricted area but within the area itself, often require additional procedural actions in order to ensure they do not alarm during normal daily activities. These types of sensors, unless used in areas personnel are not usually present, can be configured to a "bypass" mode during normal business hours and activated only upon close of business in the area where they are located. Doors to pass-through autoclaves or equipment/maintenance crawl spaces that are large enough for a human to navigate should also be secured and alarmed as appropriate.

Electronic intrusion detection devices send alarm signals to a central monitoring station where security personnel can monitor the entire security system and dispatch alarm assessment and response personnel to the location of the alarm. The area in which the central monitoring station is located should be a restricted area, and the personnel who have the authority to manage the alarms and intrusion detection configuration should be subject to appropriate personnel security measures.

3.1.2.4 Alarm Assessment

If security officials are notified about an attempt by someone to gain unauthorized access to a restricted area, the incident needs to be evaluated or assessed to determine whether the alarm is false or valid. Equipment malfunctions, accidents, and even animals can be the source of a suspected intrusion, and none of these occurrences warrant an official security incident response. Records should be kept on each actual and each false or nuisance alarm. Each record should contain the date and time of the alarm, the cause of the alarm, or a probable cause if a definite cause cannot be established, and the identity of the recorder or the operator on duty. Analysis of these records can indicate what corrective measures need to be taken to minimize the false-alarm rate.

Alarms, when generated electronically, can be assessed in person or by using remote video assessment. If an alarm is triggered, the location and nature of that alarm should be displayed at a central monitoring station. If video cameras are integrated into the security system, they should be configured to record prior, during, and immediately following the alarm event. In other words, the images should be continuously buffered in the video system memory and, if an alarm occurs, the system can retain those segments of memory that represent the images that the video camera recorded 30 seconds prior to and several minutes after the event. This configuration is considerably more efficient than a video surveillance system in which human operators are responsible for detecting security incidents and other anomalies. Properly configured video assessment systems allow an individual who is monitoring the security system to remotely assess an alarm to determine whether further investigation is necessary.

If an alarm is assessed as valid, someone trained and equipped to apprehend an intruder should respond to the alarm. If an on-site guard force, trained in this type of activity, is dispatched to assess the alarm, it may also serve as the response force. If the assessment is conducted by someone who is not equipped to handle an encounter with an intruder, that individual should summon either on-site security personnel or local law enforcement (LE) to respond. Responses to security incidents will be covered in more depth in Chapter 4.

3.1.3 Integration with Laboratory Biosafety

Controlling access to laboratories can also enhance safety by limiting the number of individuals who may be exposed to the hazards. It may be important to restrict access to a certain laboratory to only those individuals who are professionally qualified to be there. Further, in some cases, it may be necessary to limit access to individuals who have proper immunizations. The use of electronic access controls or procedural requirements to log individuals into restricted areas creates a record of who was in the area when. This information could be beneficial during investigations of laboratory safety or security incidents.

Access controls should be implemented with an awareness of laboratory operations and biosafety practices. Biometric readers can be particularly problematic. Fingerprint readers on storage containers, such as freezers, would require personnel to remove their gloves to open the container. Eye scanners may not work properly in situations where personnel are wearing eye protection, such as safety glasses, goggles, or face shields. Removing personal protective equipment (PPE) to gain access will place a burden on the normal flow of operations and may present a safety hazard. If biometric readers are desired, locating the controls at the entrance to the anteroom of a containment laboratory may be operationally feasible. Alternate access controls, such as keys, access cards, or PINs, should also be considered.

Access controls need to be put into practice in a manner that does not hinder emergency response. A mechanism must be in place that allows for the emergency entry of responders but still ensures the security of the protected materials. The mechanism may be based on procedures or systems. Procedural access for responders in emergency situations should be detailed in emergency response plans and may differ depending on the nature of the emergency. For instance, given the low fire load and with a full fire-suppression system in place, some high-containment laboratories have specifically instructed firefighters not to enter a laboratory to fight a fire. Others require emergency responders to wait for the arrival of a designated facility official to enter buildings or specific areas with MMURs and HMURs. Alternately, facilities can choose to provide authorized access to emergency responders by providing them with keys and access codes prior to an incident. This may be best suited for facilities with LMURs and/or MMURs. A key box system may be used. A key box (e.g., “Knox box”) is a small, wall-mounted safe that holds building keys for emergency responders to retrieve in emergencies. This method may be preferred over providing keys to emergency responders, since the keys might become outdated or misplaced. These boxes can provide a single-point security failure if an adversary were to obtain access to the box. To mitigate this risk, it is recommended that any key box be connected to the alarm system. Facilities with high and extreme malicious use risk agents may even consider screening or preauthorizing a small number of emergency responders (either their own or under a Memorandum of Understanding [MOU] with local responders).

Likewise, staff members must be allowed to quickly and safely exit a laboratory during an emergency. However, life-safety measures should not allow an adversary to gain unauthorized access to biological materials by activating an alarm that implements emergency egress procedures. Many life-safety codes require doors to fail safe (i.e., open) in the event of an emergency. Facilities may need to seek an exception from local building codes because doors that fail-safe represent a security vulnerability. An exception may be feasible because it affects only a small, well-defined population that can be trained on specific emergency exit procedures. An emergency exit device may be an acceptable solution. It consists of a push button, crash-out emergency hardware, or similar device, which allows personnel to override the access controls, locks, and any door interlocks. In the event of a power or electronic control failure, the battery backup should enable the system to fail in the secure (locked) condition. Operating the emergency exit device would break the circuit to the lock and allow egress. In the event of loss of both electric power and battery backup, the system should fail in the safe (unlocked) condition. The security plan should document exactly how the system has been designed to meet both safety and security requirements in the event of an emergency.

3.2 PERSONNEL SECURITY

3.2.1 Objective and Principles of Personnel Security

Personnel security is the principal security measure for addressing the Insider threat. It is fundamentally about ensuring that only trusted individuals are given authorized access to restricted areas.

Systems, such as individual badges, can be put in place to identify authorized individuals and escort those individuals who require access but have not been subjected to the same level of evaluation for personnel reliability as an authorized individual. Some level of "trust" must be established prior to allowing anyone full access to sensitive biological materials. This trust can be established through a background investigation; the depth of this investigation should vary with the level of risk that is associated with the agent that the person has access to. These types of investigations are not available in many places and, even when available, they have limitations. For example, individuals (e.g., citizens of other countries) may not have a well-documented history in the country for their backgrounds to be well-characterized by an investigation. Facilities need to use the tools available to them in screening their insider population and sometimes must think creatively about how to establish whether an individual is worthy of institutional trust in relation to the most sensitive biological materials. Performing due diligence activities prior to entrusting an individual with sensitive duties is a fundamental aspect of operating a facility in general and implementing biosecurity specifically.

3.2.2 Personnel Security Elements

3.2.2.1 Employees

Personnel screening is one part of the process for determining who at an institution should be given authorized access to restricted areas or higher-risk materials. It is also important to consider who has a need to know or needs to have access, especially because personnel screening can never be completely effective. By ensuring that members of the workforce are suitable for the positions they hold, an institution can mitigate the risk of both accidental and malicious acts. The comprehensiveness of the personnel evaluation should be commensurate with the individual's level of responsibility or position risk. Low-, moderate-, or high-risk designations can be assigned to each employment position, based upon the position's level of responsibility and access to dangerous biological agents. A standard set of personnel screening requirements (e.g., background investigations, personality tests) should be developed for each risk designation group. The screening requirements increase in rigor and intensity with increasing position risk, from low to moderate to high.

Individuals in low-risk positions generally have no contact with dangerous pathogens or toxins or do not need access to restricted areas. In general, these individuals do not have duties for which mistakes, poor judgment, or an abuse of position would cause the institution significant harm. Low-risk positions often make up the majority of positions at bioscience institutions. Little or no personnel screening may be needed for these low-risk positions.

Moderate-risk positions are those with duties that are considerably important to the institution, including those with significant program or delivery-of-service responsibilities. Examples of positions that may be considered moderate-risk include scientists and other lab personnel with direct access to MMURs, shipping and receiving personnel who handle MMURs, laboratory support personnel who require unescorted access to areas containing MMURs (e.g., safety personnel, maintenance personnel, housekeeping personnel, animal husbandry personnel), computer/network support personnel without root administrative access, and unarmed security force personnel.

High-risk positions are those positions with duties that have a broad scope of responsibility and authority. These duties are especially critical to the institution because of the potential consequences that could be incurred if the individual performed actions that were not in the interest of the institution. Examples of positions that may be considered high risk include scientists and other lab personnel with direct access to HMURs, shipping and receiving personnel who handle HMURs, laboratory support personnel who require unescorted access to areas containing HMURs (e.g., safety personnel, maintenance personnel, housekeeping personnel, animal husbandry personnel), supervisors of those in moderate- or high-risk positions, locksmiths for restricted areas, computer/network personnel with root administrative access, personnel with administrative access to the security control system, and armed security force personnel. Individuals who need to be authorized for access to EMURs may also be considered to hold high-risk positions, or the institution may determine that those individuals should be subject to more intensive or frequent background checks.

The efforts made to ensure an individual is trustworthy should be commensurate with position risk. It is often in the best interest of the institution to prequalify prospective employees for moderate- or high-risk positions prior to extending an offer of employment. Not only will time and money be saved by eliminating additional investment in security screening but some institutions may find it very difficult to release an individual from employment once the person has been hired. A basic check of qualifications and references is generally sufficient for this purpose. Some institutions may also choose to include a criminal check as well. For those institutions that hold HMURs or EMURs, it may be valuable to review as many of the easily verified elements of the full security vetting process as possible in the preemployment qualification check. These could include a verification of not only the individual's professional background but also whether he has a criminal background, financial instability, or drug or alcohol dependence. In some extreme cases, personality evaluations can be conducted. Individuals should be granted authorization to enter restricted areas without an escort only after their personnel screening is successfully completed. Institutions can make positions contingent on an employee's ability to successfully meet the personnel screening requirements.

Traditional personnel screening examines an individual's background to see if any derogatory information is uncovered. Derogatory information is unfavorable information regarding an individual that questions the individual's eligibility or continued eligibility for unescorted access authorization to restricted areas or materials. Such information that is uncovered during the personnel screening process should not necessarily disqualify someone for a position. The information can be evaluated for severity, whether it was a recent or distant event, and the frequency of occurrence. Undesirable behavior that is reported repeatedly and is recent may weigh more heavily in the evaluation, possibly raising the issue to the level of a more serious offense. Examples of information that might be considered derogatory include: association with terrorist or criminal organizations, undesirable patterns of conduct (e.g., alcoholism, drug addiction, financial irresponsibility or major liabilities, dishonesty, lack of employability for negligence, misconduct, or criminal conduct), drug-related offenses (e.g., manufacturing, trafficking, sale, or use), major honesty issues (e.g., extortion, embezzlement, or perjury), violent behavior (e.g., rape, aggravated assault, arson, child abuse, or manslaughter), illegal use of firearms or explosives, disorderly conduct, assault, criminal mischief, harassment, or employment-related misconduct (e.g., insubordination, absenteeism, or rules violations), among others.

Not all institutions will have access to resources that would facilitate personnel screening. In those situations, the institution may want to check with local LE agencies to determine if they can provide assistance. LE may be able to provide criminal histories and possibly information regarding terrorist or extremist affiliations. The institution can also speak directly to personal and work-related associates of the individual being hired into a moderate- or high-risk position. The higher the position risk, the broader the scope of the investigation should be, including the number of years in the past that are examined, the number of individuals who are questioned, the types of issues to be explored, and the frequency of reinvestigation.

The institution should have a documented rationale for how the information gathered during the personnel screening will be evaluated and used. All screening results and evaluations should be treated as sensitive information. A single offense of significant consequence may warrant a decision not to hire an individual or to remove an individual from certain duties. A lesser offense might warrant an interview that provides the individual with an opportunity to explain the circumstances under which the offense occurred, possibly providing the institution with the satisfaction that the individual is suitable for the position despite the recorded offense.

Personnel issues should not end once a person has been hired and authorized for access. In addition to regular reevaluation intervals that depend on position risk, there may be situations that warrant an immediate rescreening. For example, if an individual in a high-risk position is arrested for breaking the law, management may initiate a reevaluation. Institutions can also be proactive in helping to create a positive environment that minimizes the likelihood that personal problems may deteriorate into situations of security concern. The U.S. Secret Service has found that when a malicious act is conducted by an insider, a negative work-related event is the most common cause — the individual often has a work-related grievance and is motivated by revenge.[2] These conditions may be observable by an attentive manager or coworker who is willing to notify management, and an active Employee Assistance Program (EAP) may resolve them without incident. An EAP program is a resource for employees who may have questions or concerns about financial matters, mental health, or substance abuse. EAP programs are another mechanism that institutions can use to help ensure that employees can perform their jobs in a reliable and safe manner.

Management is responsible for ensuring that those they supervise are fit for duty. Proactive monitoring of the state of mind and health of employees will reduce the number of safety- and security-related incidents at the facility. Managers and others who are responsible for laboratory operations should intervene when an individual does not appear to be in a suitable state to work. The greater the safety or security risk, the more important it is that there are those who are empowered to temporarily remove an individual from a work environment when that individual's mental or physical health may impair safe and secure operations. Protocols for such removal or suspension should be documented in the facility's policy and be available to employees so that, should such an action become necessary, the individual will understand the basis for the action.

FIGURE 3.2 (See color figure following page 16.) Identification badges.

3.2.2.2 Employee Badges

Institutions require some mechanism for identifying which individuals have been given authorized access to which areas. For all but the smallest of facilities, it is not reasonable to expect employees to remember this information; instead, every individual should wear a badge that indicates his or her access authorization. Preferably, such badges will be designed to be difficult to replicate, will include a photograph of the employee, and will possibly have an electronic access mechanism to allow the badge to also act as a key. Employee badges can also include an institution identifier, individual's name, expiration date (visible and encoded if badge contains electronic access control capability), and color coding that indicates which areas the individual is authorized to access (Figure 3.2).

The badge should be worn above the waist with the photograph in full view. A badge should be worn on the institution's property and be required for access to restricted areas, unless the badge might compromise safety. The institution's badge should not be used for unofficial identification. Any time an individual's appearance changes significantly, a new photograph should be taken and a new badge issued. Employee badges should expire at regular intervals (e.g., 5 years) and be reissued with a new photograph. Replacement badges should also be obtained if the badge is damaged in a manner that obscures the features of the photograph or impairs an individual's ability to gain access to authorized areas or materials.

Badges with electronic access mechanisms can be encoded to allow individuals access only to those restricted areas that they are authorized to enter. The badges should be updated if access authorizations change. Authorization is contingent upon meeting the personnel screening requirements of the position, having a need to know, completing biosafety and biosecurity training, and being current on any applicable immunizations.

3.2.2.3 Visitors

The term *visitor* in this context includes any person, employee or otherwise, who does not have access to a restricted area but who has permission to enter and is therefore provided access with an authorized escort. Visitors may also include individuals from outside organizations who have official business at the institution being visited; these types of visitors may either be short-term casual visitors or longer-term working visitors.

Personal visitors, including personal friends, relatives, spouses, and children, should only be permitted in unrestricted areas during normal business hours, and such visitors should remain in the company of their hosts. The host's supervisor should be notified in the event that a personal visitor is to be on-site for more than a nominal amount of time. Casual visitors include those who are visiting the facility for business purposes but who are not involved in day-to-day operations during their visit. These individuals may go on tours, receive training, or meet with collaborators. Working visitors may fit into various categories, including any individual who is not employed by the institution but who has official business to conduct on the premises.

If a working visitor is anticipated to be on-site for more than a predetermined amount of time (e.g., 30 days) or if the working visitor requires unescorted access to restricted areas, the visitor should be screened in the same manner as an employee who holds a position of equivalent risk. The screening activities should be conducted as soon as the visitor arrives or in advance of the visitor's arrival in order to avoid the impact of long-term escorting. If a working visitor is able to demonstrate to the institution that the requirements associated with the position have already been met, the host institution may allow unescorted access at its discretion.

Visits to restricted areas should be prearranged, and a visitor should display an appropriate badge. Visits to restricted areas should be limited to official business. Visitor parking should be separated from employee parking and security personnel should be responsible for ensuring that vehicles are parked in appropriate areas. Delivery vehicles should be routed appropriately and met by receiving personnel. Drivers should either be precluded from entering restricted areas or kept under escort. All visitors should have a host at the facility being visited responsible for ensuring that all facility policies and procedures are followed and that the visitor is managed in an organized and professional manner. The host is responsible for ensuring that the visitor has completed all appropriate paperwork and is properly escorted.

The host is also responsible for ensuring that each visitor is issued an appropriate badge. A visitor's badge may or may not include a picture; if the visitor is a working visitor, however, a picture badge similar to those issued to employees but with an indication that it is a visitor's badge is preferable. Visitor badges should expire upon termination of the visit or at a standard interval (such as annually), whichever comes first.

Individuals who have not been given authorized access should be escorted. This includes visitors; support personnel who are required to enter the area for maintenance, repairs, or cleaning but who are not cleared for access; and all other individuals without the appropriate institutional identification and keys that would provide them access to a specific area. Those individuals who have not been fully vetted should also remain under escort in restricted areas until their screening is successfully completed and evaluated. It is important to establish clear escort policies and procedures, such as appropriate visitor-to-escort ratios, responsibilities of the escort or host, and rules for after-hours visitors.

A visitor to a restricted area should fill out a visitor log or have identifying information logged into the area electronically. The visitor log should include the names of the visitor and the escort, their signatures, the visitor's organization, the purpose of the visit, badge number (if applicable), and the times at which the visitor entered and exited the area. Each restricted area should have an actively maintained a visitor log that is in chronological order for a year or more; the log should then be subsequently archived. Those who are not authorized for routine access, but who have legitimate business in the area, should be escorted by an authorized individual and sign a visitor log.

3.2.3 Integration with Laboratory Biosafety

The basic principles of personnel security have many benefits to laboratory biosafety and are often already in place at institutions with strong biosafety programs. It is prudent practice to verify a person's technical background and training before giving the individual access to a laboratory. Such basic screening gives an institution a degree of confidence that personnel can be trusted to work safely with specific biological agents. Institutions typically have more stringent requirements for experience and training before giving personnel access to enter areas that present a higher biosafety risk. Likewise, escorting visitors in laboratories helps ensure their safety.

Identification badges are often used to indicate authorized access and sometimes to enable entry, but they are not compatible with the operating reality of many containment laboratories. In situations where wearing a badge would result in compromised safety, another mechanism for indicating authorized access can be selected, such as a keypad for PIN entry. Another possible solution is to provide a locker or other secure storage space in the anteroom for keys, badges, and other personal belongings. Keys, access cards, and other such devices also present a potential contamination issue in containment laboratories. Do keys need to be decontaminated before removal from the laboratory?

3.3 MATERIAL CONTROL AND ACCOUNTABILITY

3.3.1 Objective and Principles of MC&A

The objective of MC&A measures is to create an environment that discourages insiders from stealing and using biological agents maliciously. MC&A seeks to adopt practices that establish and reinforce responsible oversight of work with dangerous pathogens and toxins.

MC&A measures help enhance laboratory biosecurity by establishing exactly what biological *material* is present at a facility, how and where the material is stored and handled, and who is responsible for it. MC&A combines policies, procedures, and technologies to augment other elements of laboratory biosecurity during the use, storage, and transfer of material. *Control* ensures that material is confined to known, legitimate use, whereas *accountability* ensures oversight by formally associating material with people and information records. Despite the fact that it is not possible to count every microbe in the laboratory environment, it is possible to take prudent measures to ensure that dangerous pathogens and toxins are controlled in a manner that will deter, and possibly detect, theft of these materials. At a minimum, MC&A measures can facilitate forensic analysis if an illicit diversion is detected.

3.3.2 MC&A Elements

3.3.2.1 Material

The first aspect of MC&A is deciding which materials are subject to control and accountability measures. This decision requires identifying the agents, the form of the agents, and whether quantity is a factor. Ideally, material subject to MC&A should be identified through a rigorous risk assessment; there may also be regulatory requirements that define these materials specifically.

The facility risk assessment (as described in Chapter 2) should identify and categorize those materials that require MC&A measures. In addition to MMUR, HMUR, or EMUR agents, consideration should be given to those subcomponents, any special experimental form, or other variant deemed to present a commensurate risk. MMUR, HMUR, or EMUR agents and their controlled subcomponents and variants are herein after referred to collectively as "dangerous biological agents."

The difficulty in defining "material" subject to MC&A is in the details. Dangerous biological agents can be found in many specimens throughout a typical bioscience institute: repository stock cultures, working stocks, clinical specimens, unknown samples, and genomic material, among other categories. These specimens are also found in many forms: liquid solution, lyophilized powder, in an infected host, in animal waste, and as contamination on equipment or other objects. Genomic libraries of dangerous pathogens and identified virulence genes from an organism may also be considered material from an MC&A perspective. It may not be important to include samples of unknown content unless they are suspected of containing dangerous biological agents. Of course, once a sample has been positively identified as a dangerous biological agent, it should be subject to specific MC&A measures that are appropriate to the agent. It must be clear where specific MC&A measures begin and end for specific agents. The manner in which MC&A is implemented may vary for different types and forms of specimens or agents.

MC&A does not apply to equipment, instruments, clothing, and similar laboratory objects that have been, or may have been, contaminated with materials. These items should be decontaminated and, if necessary, disposed of properly in an appropriate area, foregoing the need to include these items in MC&A procedures.

3.3.2.2 Control

Control is implemented to ensure that materials stay where intended and that they are used for a stated purpose by specifically designated and authorized people. Control must encompass all activities involving the material, such as storage, use, transport (see Section 3.4), and disposal. Control should be effective under both normal conditions and anticipated abnormal conditions wherever possible, such as accidents, power failures, or emergencies. Otherwise, covert diversion of material could be attempted under the cover of an abnormal condition (perhaps intentionally caused).

Control can be accomplished in one of two ways. Physical control is a means of preventing unauthorized access, such as locking a freezer or limiting research with the materials to a restricted area. Control can also be accomplished through operational procedures. Physical and procedural control measures aim to (1) assure the integrity of each material item, (2) assure that no item is missing, and (3) minimize the opportunity for misuse during activities that necessarily involve bulk material.

Storing and using dangerous biological agents in restricted areas and limiting access to these materials to authorized personnel help establish control. A locked freezer or vault can offer additional control in laboratory spaces that are shared among more individuals than actually work with a particular material. Automated systems can log events such as laboratory access (in and out) or freezer opening and closing, often providing date, time, and personal identification. Barcode labels or radio-frequency tags are engineered measures that can facilitate inventory taking. Another physical control mechanism is item integrity, which means establishing a complete boundary around an item so that material within the item cannot easily cross that boundary without detection. Item integrity is usually not practical for working stocks, but it may be valuable for higher-risk repository stocks. It can entail the use of an enclosure ("seal") that would reveal any attempt to tamper with the item. Seals require inspection and verification that should occur periodically, or before intended, legitimate use.

There is often a balance between physical and procedural controls. For example, if the required autoclaves, incubators, centrifuges, or other specialized equipment are not located within the same restricted area, procedural controls can be implemented to ensure materials are under the same standard of control throughout their life cycle.

Procedures, designed in advance and explicitly considering their laboratory biosecurity implications, are central to the effective control of pathogens and toxins. MC&A procedures can typically be integrated with current standard laboratory procedures. MC&A elements may be appropriate to include in many procedures, such as working with the material, inactivating and disposing of the material, conducting inventory checks, labeling sample containers, removing material from storage, and returning it to storage. In addition to procedures for routine activities, unexpected conditions should be anticipated wherever possible. What is the course of action if a sample is discovered missing?

Control is greatly facilitated if material is aggregated into *items*, i.e., discrete, identifiable, and countable units. Material in solution in a test tube or sealed in an ampoule can be aggregated as an item, for example. Similarly, material in an animal carcass might be considered an item, but it would be better if the carcass itself were confined in a sealed box; the box would then become the item. It may not be necessary to define items at the most detailed level possible. An item could be a cabinet or freezer, or even a restricted-access laboratory, provided that control measures can assure item integrity. If it can be done, such higher-level aggregation would obviate any need to count vials.

Regardless of how it is defined, each item should correspond one-to-one with an associated information record for accountability purposes. Although the associated quantity of material within the item can sometimes be specified, it is essentially irrelevant for replicating organisms; any amount is significant. Instead, the number of containers or accountable *items* is the relevant quantity to track for replicating organisms. The quantity of toxins and other nonreplicating materials is meaningful, however. Because the risk for nonreplicating dangerous biological agents is quantity-dependent, the institution can only accurately characterize its risk if it is aware of the total quantity of material at their facility. This implies the necessity of some basic level of MC&A measures for nonreplicating materials, with the sophistication of the controls and accountability increasing with the quantity of material.

3.3.2.3 Accountability

Accountability is the means of ensuring that someone is responsible for the dangerous biological agents stored and/or used within a defined area. Assigning qualified, authorized individuals to oversee the control of protected agents, keeping accurate and timely records, reporting, and auditing are all aspects of accountability.

Each dangerous pathogen and toxin should have a designated "accountable" individual who is knowledgeable about the assigned pathogen or toxin in storage and in use. An accountable individual may be assigned on an agent-by-agent basis, on a per-laboratory basis, or using any other convenient distinction. The critical characteristic of this individual's accountability function is the responsibility of having an ongoing awareness of an agent's status within the laboratory. The accountable individual is responsible for providing information about how, when, where, and why assigned pathogens and toxins have been used, transported, or destroyed, and for maintaining current accountability (inventory) records. Accountable individuals are responsible for overseeing the work associated with their assigned pathogens or toxins. Any anomalies noted by the accountable individual should be reported to the appropriate officials promptly. The head of the facility should be responsible for ensuring that an appropriate accountable individual has been assigned to each dangerous pathogen or toxin located in the facility.

Most records exist for the institution to retain historical knowledge about MC&A-subject material. Exactly how such knowledge is recorded may vary considerably according to the particular circumstances. The objective is to describe both the existence and use of material accurately in a timely manner, and completely, so that an accountable person can answer questions that could arise later. Laboratory notebooks, inventories as manual ledgers or electronic files and databases, and shipping/receiving receipts are all examples of material records that contribute to accountability. To serve these functions, the information records should unambiguously indicate the specific material and the associated accountable person.

Decisions must be made about several aspects of record keeping:

- *When* must information be recorded?
- *What* information must be kept?
- *How long* must information be kept?
- *Where* should information records be kept?
- What *security* is required for these records?

There is a variety of information that is important to document. First, the attributes of the material must be captured to characterize the material, i.e., to describe what it is. This category includes the agent strain information and possibly its origin, date of acquisition, source history, quantity, and various measured data. A description of the item is necessary to identify which item it is. When multiple items exist, it is especially important to specify the container, identifying information from the label, and its location. It may also be useful to document the status of the material (i.e., active working samples in use, only stored seed stocks exist, material destroyed or transferred to another accountable person or institution) and the associated dates for any relevant change in status. The records should provide the name and contact information of the accountable person for each material. The inventory should include all biological materials that are subject to MC&A measures at the facility. The inventory should cover all repository stocks as well as any unique moderate-, high-, or extreme-risk isolates. The facility inventory should include information about the location of each of the dangerous pathogens and toxins and its associated accountable individual. For higher-risk agents, a certain subset of working samples may also be subject to formal inventory controls.

A physical inventory is used to reconcile these records (the "book inventory") with the materials that are actually at the institution. A physical inventory is accomplished by identifying and listing all subject materials item-by-item in a particular area, such as a laboratory or a workstation within a laboratory. The physical inventory is assembled by a thorough search and review of all locations where the subject materials may exist. A physical inventory should be conducted periodically, with the frequency depending on the agents involved; the higher the risk, the more frequently physical inventories should be conducted. Whenever a physical inventory is conducted, the results should be compared with the current book inventory, and any discrepancies should be identified. If a discrepancy indicates the possibility of theft, or if the discrepancy remains otherwise unexplained, it should be reported immediately to the accountable individual and, if appropriate, local and national authorities. Conducting a physical inventory includes identifying items, counting items, and occasionally conducting tests to verify contents. When large numbers of items are involved, especially for lower-risk materials, selective (statistical) sampling for identification or diagnosis may be employed.

MC&A information might prove useful to an adversary, so it should be treated as sensitive information and should be subject to information security practices. MC&A information is often intermingled with information recorded for scientific purposes, so care should be taken to prevent sensitive MC&A information from inadvertently being released to the public. A detailed inventory should be kept in a secure, limited-access database.

3.3.3 Integration with Laboratory Biosafety

Measures may already be implemented in various institutions that appear to be (and may in fact be called) MC&A. From a laboratory biosafety perspective, it is important to know what materials are present at an institution, which ones are in active use, which ones are just held in storage, who uses the material, and who is responsible for it. These are the same principles that underlie MC&A for laboratory biosecurity. However, when MC&A measures are already in place at a facility, they should be reviewed explicitly from a laboratory biosecurity perspective and updated as appropriate. New procedures specific to biosecurity should be developed where gaps are identified.

The CDC and NIH *Biosafety in Microbiological and Biomedical Laboratories* (*BMBL*) and the WHO's *Laboratory Biosafety Manual* both require a biohazard sign for laboratories at BSL2 or higher. This sign is intended to provide notification of potential biohazards, such as the specific biological agents, present in the room. The biohazard signs normally include the name of the agent, specific hazards, and contact information of the investigator. Identifying the agent, its location, and the name of those individuals responsible for that agent may conflict with the objectives of biosecurity. Depending on the location of the sign, it can identify the location of biological agents to those who do not have a legitimate need to know that information. The locations of biohazard signs for laboratories with MMURs or higher must be carefully planned to avoid compromising security while providing the necessary level of safety. The first consideration is to assess who has access to the laboratory door where the sign typically would be placed. Is the laboratory in a restricted corridor that can only be accessed by authorized individuals? Does the laboratory have an anteroom? If so, a simple biohazard sign can be placed on the door to enter the restricted access anteroom, and biohazard sign that lists the specific biological agents can be placed on the inner door. A less desirable but still plausible solution may be to post a simple biohazard sign on the laboratory door and place specific information on the biological agents and other hazards in a designated location right inside the laboratory entrance. Personnel with access to the laboratory must then be trained on the location of this information.

3.4 TRANSPORT SECURITY

3.4.1 Objective and Principles of Transport Security

Transport security is a mechanism to implement MC&A to reduce the risks of insider and outsider theft while material is being transported between restricted areas. This transport can be within a facility, between facilities within a country, or internationally.

Scientists, health agencies, and diagnostic laboratories rely on the timely exchange of biological materials for a variety of reasons. During the process of transportation, materials move outside of established restricted areas and may be more vulnerable to theft or tampering. Accountability of the material, documentation, and oversight during the transport process are measures that improve biosecurity. Transport security can reduce the likelihood of (1) inappropriate handling and movement of material by scientists or technical staff, (2) the possibility of loss or misplacement of material during transfer, and (3) the possibility of theft of material for malicious purposes, which may indicate vulnerabilities for terrorist activities as well.

3.4.2 Transport Security Elements

Internal transport is the movement of dangerous pathogens and toxins between restricted areas at a facility. External transport refers to the process of moving dangerous pathogens or toxins between facilities (Figure 3.3).

3.4.2.1 Internal Transport

The usual internal transport process is straightforward. Typically, an individual in the originating laboratory removes a sample from storage, walks it across the facility, and hands the sample to an individual in the receiving location. The receiving laboratory either uses the sample or places it in storage. Such movement may occur as laboratories exchange materials under study; internal transport also

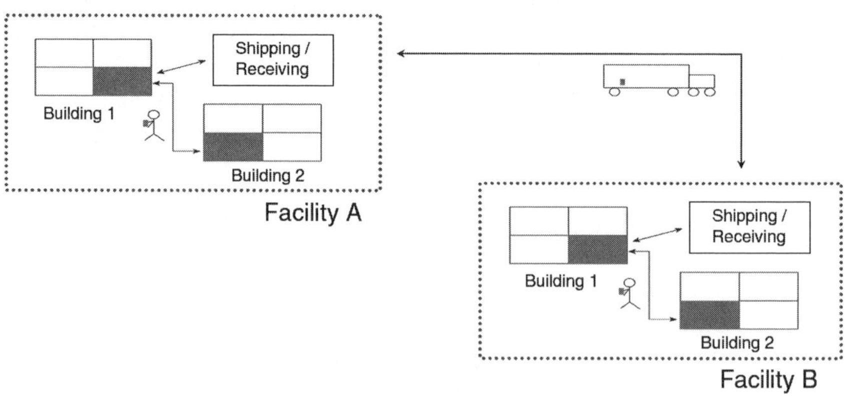

FIGURE 3.3 Transport.

includes materials added or removed from the inventory as a result of shipping and receiving processes, or materials sent to disposal areas (e.g., autoclave and incinerator rooms). Either laboratory or shipping personnel may be responsible for preparing packages in accordance with all appropriate regulations.

Because materials may be vulnerable to theft while outside restricted areas, facilities need to demonstrate prudent and sufficient stewardship of these materials during transport, with more stringent measures in place for the transport of higher-risk materials. A key problem for internal transport is determining at which point a diagnostic sample should be subject to internal or intrafacility transport security procedures. Are the measures implemented upon arrival of a specimen at the facility if it is suspected of containing a dangerous biological agent, or are they initiated after the sample is determined through confirmatory testing to be an agent that requires transport security?

The same basic principles of MC&A are applicable to internal transport security. Everyone who has physical access to dangerous biological agents in transport should be subject to the same personnel security requirements as those required for other individuals with access to the materials in the laboratory. Facilities should recognize that laboratory personnel are not necessarily the only ones with unescorted access to dangerous pathogens and toxins during transport. Transport protocols can be analyzed to determine who may have access to such materials, such as dedicated delivery people. Analysis of transport processes may identify areas that are used for temporary storage, such as shipping and receiving offices or loading docks. Controls should be implemented in these areas at a level equivalent to the restricted areas where the material is stored or used.

Transport of materials should also be integrated into the MC&A protocols. The accountable individual for the material preapproves the transport. For higher-risk materials, a facility may decide that transport must be preapproved by a designated institutional responsible official or biosafety officer (BSO). Prior to approving the transport, the individual ensures that a new accountable individual is identified at the recipient laboratory or verifies that appropriate shipping or destruction documentation is maintained.

Chain of custody refers to the process employed to document who has control of a sample and when. The chain-of-custody process documents that an accountable individual has control over the integrity of the packaged material, and that secure receipt of the material has occurred at the appropriate facility location. Chain-of-custody documentation accompanies the material during transport and includes the name and quantity of material being moved, the shipping and consignee contact information (or laboratory contact information as applicable), and time and date signatures of every individual who assumes control of the material *en route* (e.g., those who initiate delivery, package, or relinquish custody). The chain-of-custody process also documents any situation in which an individual assumes custody on behalf of another individual. If an authorized individual is not able to ensure custody of the package, then the package can be controlled in a restricted area or within an access-controlled cage or freezer. Chain of custody can be achieved via many mechanisms, including paper, where each individual signs a physical document, or personal digital assistant (PDA) scanners. Chain of custody does not guarantee that a sample will not be stolen. However, it does raise the threshold by introducing a degree of accountability in the transfer process.

3.4.2.2 External Transport

This process likely includes internal transport steps (e.g., to a shipping area or from a receiving area) in addition to relinquishing custody to a commercial carrier or courier. A facility cannot guarantee or oversee the security of material outside of the facility, but consideration of external security issues can limit the possibility of incidents. As warranted by the risk, additional procedural steps can be taken by shipping and receiving facilities to exercise due diligence during all three stages of external transport: preshipment, *en route*, and receipt.

The facility can require laboratorians to obtain preshipment authorization. The authority to approve shipment may change depending on the risk of the material being transported. For lower-risk agents, the accountable individual could be empowered to make shipping decisions, whereas approval from a designated facility representative could be required prior to shipping a higher-risk material. As an element of the preapproval process, the sending laboratory should have knowledge of the professional capabilities of the receiving laboratory. The originating laboratory can notify the receiving laboratory of any shipping-related information (e.g., tracking number, time of shipment, expected time of receipt).

Security while the package is *en route* begins with the selection of a reputable carrier. For higher-risk materials, carriers with transportation security plans can be selected. The opportunity for theft is greatly reduced by limiting exposure time to the commercial transport system. Rapid shipments via air reduce exposure time. Air services may also have well-controlled staging and bulk break areas restricted to authorized employees. Stealing dangerous pathogens or toxins from a commercial overnight carrier would be more difficult than other possible forms of transportation. The packaging should not attract any special attention; labels on the outside packaging should have only the minimum identifying information required by the commercial carrier.

Some air freight service providers provide a *constant surveillance* service. This service can extend the time until delivery and increase the cost of transportation. Because of the additional transportation time, constant surveillance is *not* recommended. Although many commercial carriers provide tracking services, it should be recognized that these services are not in real time, nor do they guarantee custody over a package at all times. However, tracking does provide information regarding the relative position of the package in the transport system and can facilitate creating a document trail for facilities. Facilities can designate individuals who are responsible for package tracking and monitoring during the external transport.

The receiving facility can provide notification to the sending facility that the material has been successfully received. Both laboratories can be prepared to independently and immediately notify the shipping company if any shipment does not arrive as expected. There should also be procedures for reporting missing shipments of higher-risk agents to appropriate authorities.

Procedures should also be established so that an accountable institutional official is notified in advance that a shipment of dangerous biological materials will be received at the facility. That institutional official can ensure that only authorized individuals receive custody of that particular package when it arrives at the facility, and that it is rapidly delivered to the appropriate restricted area within the facility. Procedures should also be created to accommodate those circumstances when a package of dangerous biological materials arrives at a facility without prior notification or without clear indication of the intended recipient.

3.4.3 Integration with Laboratory Biosafety

Transport security mechanisms must coexist with a large body of safety regulations; must allow for the efficient transportation of all materials, especially frozen materials; and must remain cost effective so as not to unduly hinder the research and diagnostic work that is essential for advancing public and agricultural health.

Limited access to dangerous pathogens and toxins during transport can be complementary to transport safety issues. Restricted access means fewer people to train and less exposure risk in the advent of a spill.

Under safety regulations, the maximum amount of dangerous animal or human pathogens or toxins that may be transferred is 50 ml liquid/50 mg solid by passenger airplanes, and 4 liters liquid/4 kilograms solid by cargo airplanes. Quantity limitations also provide some security benefits. The opportunity for theft is also reduced by limiting the amount of time that the biological agents are outside of restricted areas.

3.5 INFORMATION SECURITY

3.5.1 Objective and Principles of Information Security

Information security is a set of tools and practices used to protect sensitive information. Protecting sensitive information from release is a security measure because release of this information could aid an individual's efforts to steal protected biological agents by indicating how to circumvent the laboratory biosecurity system.

This section is devoted to protecting information that may be considered sensitive, particularly sensitive security-related information. However, other forms of sensitive information, such as personnel and financial records, may also warrant these types of protections, including legitimate restriction from public access. Information, such as experimental data and proprietary information which may be considered a valuable asset, should be held redundantly by the institution, thereby reducing the consequences of loss to a negligible amount and minimizing the risk of theft or sabotage.

3.5.2 Information Security Elements

3.5.2.1 Sensitive Information

The first step in providing information security is identifying information that is sensitive. In this context, sensitive information could help an adversary circumvent the security system to acquire protected biological agents. This type of information not only should be protected from public disclosure but also should be limited to specific authorized individuals. Sensitive security information may be connected to any of the elements of laboratory biosecurity. All physical security information warrants some level of protection. Information regarding physical security plans, user-level access, or other details of the physical security system is sensitive. Similarly, facility plans, including blueprints and other details, may be considered sensitive and need to be protected. Other sensitive physical security information includes physical security system manuals, passwords, and other system-specific details. Laboratory notebooks with MC&A records; material inventories, whether electronic or paper; transportation documentation and similar records may also be sensitive.

In addition to representing an employee confidentiality concern, sensitive personal information can be a biosecurity concern. Background investigation and personality test results could be used in an inappropriate manner to coerce or embarrass a person with access to dangerous pathogens or toxins. Additionally, personnel access authorizations could lead an adversary to target a particular individual for collusion or to monitor that person in the hopes of learning some specific security procedures that may be in place.

A process should be established for review and approval of all potentially sensitive information prior to public release. By implementing such a process, sensitive information can be identified and protected from inadvertent disclosure. Sometimes information can be modified in a way that it is no longer sensitive. Aggregate summaries of inventory information, for instance, may not be sensitive. Generalities regarding risk or security may also be acceptable for public consumption. The review and approval process can be designed to help identify how information can be modified to make it appropriate to release publicly.

Once sensitive information is identified, appropriate security measures can be designed and implemented. These measures encompass handling, storing, transmitting, and destroying such information. The same fundamental principle of limiting access to authorized individuals is central to information security. In addition to personnel screening, authorization requires determining who has a need to know. The owner or originator of the sensitive information should determine who else needs access to that information to execute professional responsibilities. In addition to this need to know, those persons should be authorized to have access to that sensitive information. In most cases, authorization should be dependent upon successfully completing an appropriate amount of personnel screening to determine that the person can be trusted with that sensitive information.

Marking information that has been determined to be sensitive information helps indicate that it requires appropriate protection. The manner in which information is marked depends on the form of the information. For instance, documents may have a cover page identifying them as sensitive information. The top and bottom of each page in a file can be labeled as sensitive. Labels for removable electronic media need to be clearly visible and applied in a way that they do not interfere with the drive mechanisms. Removable electronic media include CDs, DVDs, pen drives, floppy disks, digital tape cassettes, removable hard drives, and any other device on which data can be stored, and that is normally removable from the system by the user or operator.

Sensitive information, both in hard copy and electronic form, should be physically protected and stored within a facility's restricted areas. Institutions need to consider when they will allow sensitive information to be stored and handled outside of their restricted areas. It may be permissible for employees to work at home or on travel if there is a reasonable expectation of privacy and the employee takes measures equivalent to protecting valuable personal property.

Most sensitive information will need to be shared or exchanged between authorized individuals at some point. Facilities will need to determine and institute the appropriate security protocols for transmitting or sharing sensitive information. Sufficiently secure information transfer mechanisms may include in-person discussions, use of telephone landlines (noncellular phones), fax, reputable mail services, restricted access computer networks, or encrypted and authenticated e-mail.

Destruction of sensitive information also needs to be considered. Comprehensive destruction may be warranted. For paper documents, this includes shredding or burning. A hard-drive wiping program provides a similar level of destruction for electronically stored sensitive information. Electronic storage devices can be destroyed through physical damage to the point of inoperability via shredding, degaussing, melting, or other such methods before disposal.

3.5.2.2 Electronic Information

Sensitive information may be found in many electronic formats on stand-alone computers and computer networks throughout the facility. It may be appropriate to store information that is critical to security (e.g., physical security systems, dangerous pathogen inventories) on stand-alone computers or isolated networks within restricted areas to limit the risk of compromise. All elements of the network (routers, servers, Web servers, Web applications, domains, firewalls, wireless local area networks, and remote access) need to be assessed from a security perspective. Strong passwords, desktop management of upgrades and patches, and virus protection are all important aspects of desktop security measures for any computer with sensitive information or with access to a network that contains sensitive information.

Any individual with root administrative access to the administrative or security network needs to be aware of information sensitivity levels and cognizant of any actions taken in the handling and protection of that information. Additional training and policy controls should be implemented for individuals with root access to these systems. Individuals with root access should be screened to the highest level associated with the information or controls that root access provides to them.

3.5.3 Integration with Laboratory Biosafety

MC&A-relevant information is *not* the same as laboratory research results or the scientific content of publications. Classification or limiting the public release of research findings, methods, and techniques is an entirely separate issue. The laboratory biosecurity considerations here are limited specifically to avoiding the release of information about (1) where particular dangerous biological material is and how one might obtain it for malicious use, and (2) what specific laboratory biosecurity measures are in place at specific facilities to protect dangerous biological material. Publications should be reviewed to prevent unnecessary release of such MC&A-relevant information.

Having a clear process for determining what information can be shared with whom can be a significant benefit to laboratory biosafety and the institution as a whole. A well-defined review and approval process can help ensure that the institution does not unduly restrict information from public dissemination. In some cases, without such a process, employees can self-censor information because they do not know what is acceptable to release from a legal and institutional perspective.

References

1. Garcia, M.L., 2001, *The Design and Evaluation of Physical Protection Systems,* Butterworth-Heinemann, Boston, MA; Gregg, M. and Kim, D., 2005, *Inside Network Security Assessment: Guarding Your IT Infrastructure,* Sams Publishing, Indianapolis, IN.
2. U.S. Secret Service and CERT Coordination Center/SEI, May 2005, *Insider Threat Study: Computer System Sabotage in Critical Infrastructure Sectors*, Washington, D.C., p. 22 [by Michelle Keeney, J.D., Ph.D., Eileen Kowalski at the National Threat Assessment Center, and Dawn Cappelli, Andrew Moore, Timothy Shimeall, Stephanie Rogers at the Software Engineering Institute].

4 Program Management

Program management guides and oversees the implementation of the laboratory biosecurity program and is central to its success. It is the responsibility of management to ensure that each component of the laboratory biosecurity system functions optimally and in a coordinated and consistent manner. To achieve this end, the management should identify and prioritize program needs and allocate appropriate resources to meet those needs.

Program managers should have extensive familiarity with the risk assessment and risk management process. Management should take the initiative to work closely with those personnel conducting the risk assessment and engage in open discussions about the particular assets and threats at the facility. Ultimately, after the risk assessment has been conducted, it is the responsibility of management to decide which risks must be mitigated and to allocate resources in a graded manner to meet those needs. Once such decisions have been made, it is imperative that management create a laboratory biosecurity plan, incident response plan, and any other documents (e.g., standard operating procedures [SOP]) necessary for the successful operation of the laboratory biosecurity system.

Such a plan helps ensure the overall functioning of the laboratory biosecurity program. The laboratory biosecurity plan is the central document of the laboratory biosecurity program and should outline the security measures that are implemented at the facility. The plan should offer comprehensive guidance on the implementation of laboratory biosecurity at the facility and should address the policies and reference procedures associated with personnel management, physical security, MC&A, transport security, and information security.

In the biosecurity plan, management should also clearly delineate the roles and responsibilities of laboratory personnel with regard to laboratory biosecurity. Attention should be given to ensure that all of the required activities described in the laboratory biosecurity plan can be performed effectively by the individuals who have been given those responsibilities.

To ensure that personnel are familiar with laboratory biosecurity, a variety of training programs should be implemented. These training sessions should inform personnel of their requirements regarding laboratory biosecurity. For optimal awareness, employees should be briefed upon hiring and termination, and should also be provided refresher courses throughout their terms of employment. Because different personnel have different duties, special training sessions may be necessary for managers, scientists, response force officers, and other employees.

Management must also ensure continual improvement of the laboratory biosecurity system. To do so, management should conduct routine self-assessments of the system; subsequent management reviews should evaluate the findings of these assessments and propose actions to redress significant weaknesses. Actions should be both corrective and preventive in nature — that is, they should correct existing problems and anticipate and correct for new problems that may arise.

At most facilities, program management is the overarching component of both laboratory biosecurity and laboratory biosafety. Through activities such as planning, staffing, funding, and training, program management addresses every element of the laboratory biosecurity and laboratory biosafety program. (Figure 4.1)

The following sections provide specific guidance on the management of the laboratory biosecurity program.

4.1 ROLES AND RESPONSIBILITIES

An institute's personnel are instrumental to achieving an effective, functioning, laboratory biosecurity system. Thus, it is important for management to outline individual laboratory biosecurity responsibilities.

The roles and responsibilities described in this section represent one approach. Facility-specific considerations — such as a large facility or extensive research on certain dangerous biological agents — may make it appropriate to create additional positions subsidiary to those listed. Alternately, roles and responsibilities may be consolidated in small institutions.

Management must establish the objectives of the biosecurity system, communicate those objectives to the appropriate stakeholders, and provide the necessary institutional support and resources to implement them. The director (or head) of the institution is responsible for ensuring that effective laboratory biosecurity is implemented and that incident response is in place. The director may choose to delegate specific aspects of these roles to other administrative staff. This should be noted in written documents. At large institutions or institutions with high malicious use risk agents, the director should designate a laboratory biosecurity officer.

The laboratory biosecurity officer develops the laboratory biosecurity plan. The officer has the responsibility to work with laboratory managers and laboratory staff to ensure that the laboratory is adhering to policies on laboratory biosecurity. If the laboratory biosecurity officer roles and responsibilities are not given to the laboratory biosafety officer as additional duties, these two officers must coordinate to resolve any potential conflicts between the two programs and explore opportunities to leverage resources to achieve complementary program objectives. The laboratory biosecurity officer should oversee the implementation of physical and personnel security at the institution and ensure that the laboratory biosecurity system is accurately reflected in an institution's overall site security efforts. The biosecurity program needs to be assessed, and metrics measuring the program's success need to be developed and tracked. Thus, this individual should conduct self-assessments and be the point of contact for outside assessments. The laboratory biosecurity officer should establish, develop, maintain, and update

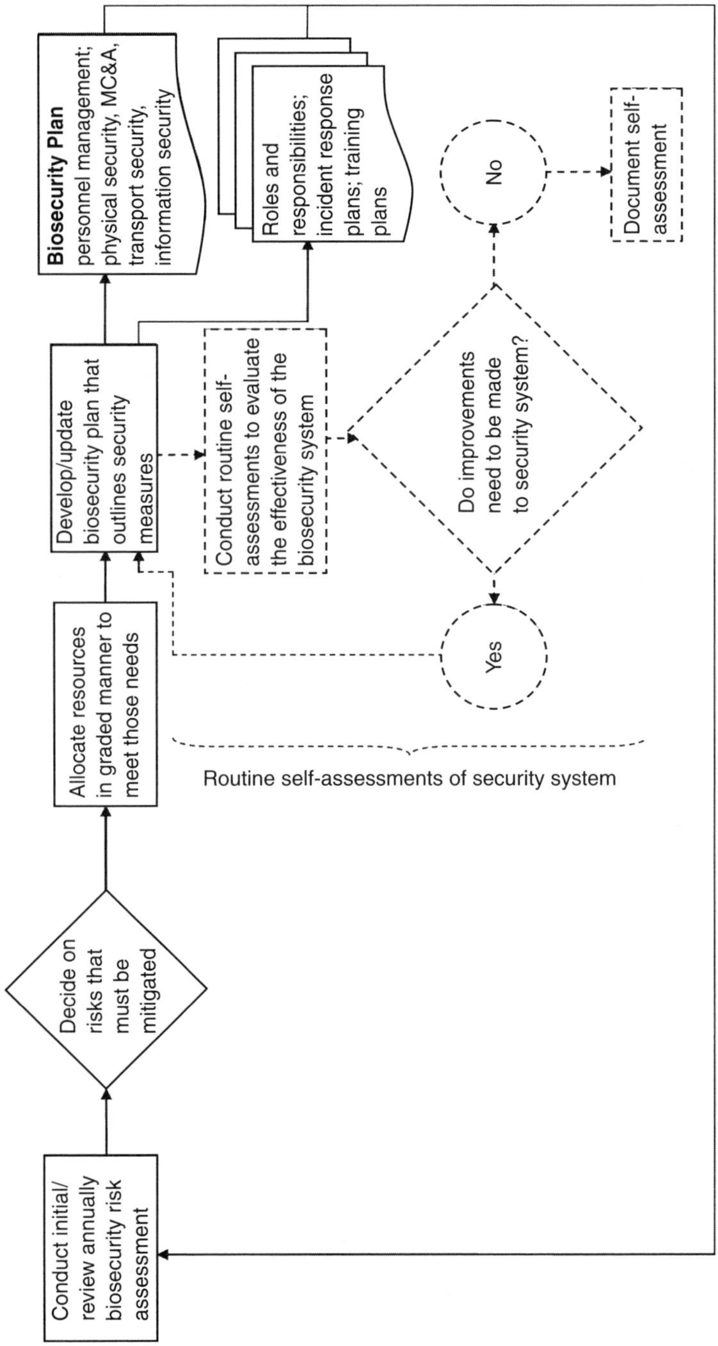

FIGURE 4.1 Program management flow the chart.

criteria for identifying and analyzing trends in laboratory biosecurity violations and other lapses in achieving laboratory biosecurity objectives and goals.

The institution needs to designate responsibility for a variety of administrative biosecurity tasks, such as personnel security management and information security. An individual should coordinate required background investigations and adjudication, oversee badging, and implement visitor requirements. Similarly, a specific individual should serve as the main point of contact for information security issues, implement information security policies and participate in the development of laboratory biosecurity plans, monitor local computer security activities, respond to computer security incidents, and ensure that there is adequate understanding of computer security policies and procedures. This individual should also coordinate information security self-assessments. Because large institutions typically already have personnel offices and information security administrators, many of these roles and responsibilities may already be assigned; the laboratory biosecurity officer or another administrative representative needs to simply review existing systems to ensure that appropriate aspects of the biosecurity system are incorporated.

An individual should be delegated the responsibility for ensuring that all biosecurity documents are properly maintained and controlled. It may be appropriate to create a central catalog of all laboratory biosecurity documents and the associated accountable individuals (analogous to an inventory of high-risk pathogens). The individual with oversight for all documents should develop schedules that specify how long documents need to be retained and set forth requirements for review and revision. There may be specific regulatory requirements for retention of certain documents. The individual then needs to ensure that documents are reviewed, revised, and retained according to this schedule. The individual will need to coordinate with the information security specialist in the development of proper procedures for document destruction and on general information security issues. Depending on the size and scope of a biosecurity program, this may become an additional duty of the laboratory biosecurity officer.

Laboratory managers (e.g., Principal Investigator [PI]) or other line management with direct oversight of researchers, diagnosticians, animal technicians, and any other individuals with access to moderate and/or high malicious use risk agents have many responsibilities for laboratory biosecurity. In coordination with the laboratory biosecurity officer, managers must implement operational laboratory biosecurity programs. Managers need to provide resources for training, implementation, and monitoring of laboratory biosecurity policies and programs. Managers should ensure that personnel receive all required training, including annual comprehensive laboratory biosecurity training. Managers should give new staff an orientation briefing. They should also meet with departing staff during the termination process to make sure that all outstanding biosecurity issues have been addressed (e.g., keys and appropriate documents have been turned in). Because of their day-to-day involvement with the laboratory, managers are also the best candidates to be trained to perform review and approval of information for public release. Ultimately, the manager's most important responsibility is to instill and maintain continued awareness of laboratory biosecurity requirements and the importance of laboratory biosecurity.

If an on-site guard force is present, it has responsibility for many routine security functions, such as limiting access at entry gates or entrances to restricted areas to authorized vehicles and personnel, monitoring parking areas and the overall site to deter illicit acts as well as to ensure that no unauthorized vehicles or personnel are present, and monitoring and assessing alarms produced by any electronic intrusion detection system. In addition, the guard force (or other on-site response personnel) will respond to any potential laboratory biosecurity incidents.

All personnel are responsible for understanding and complying with all laboratory biosecurity policies and procedures (Figure 4.2).

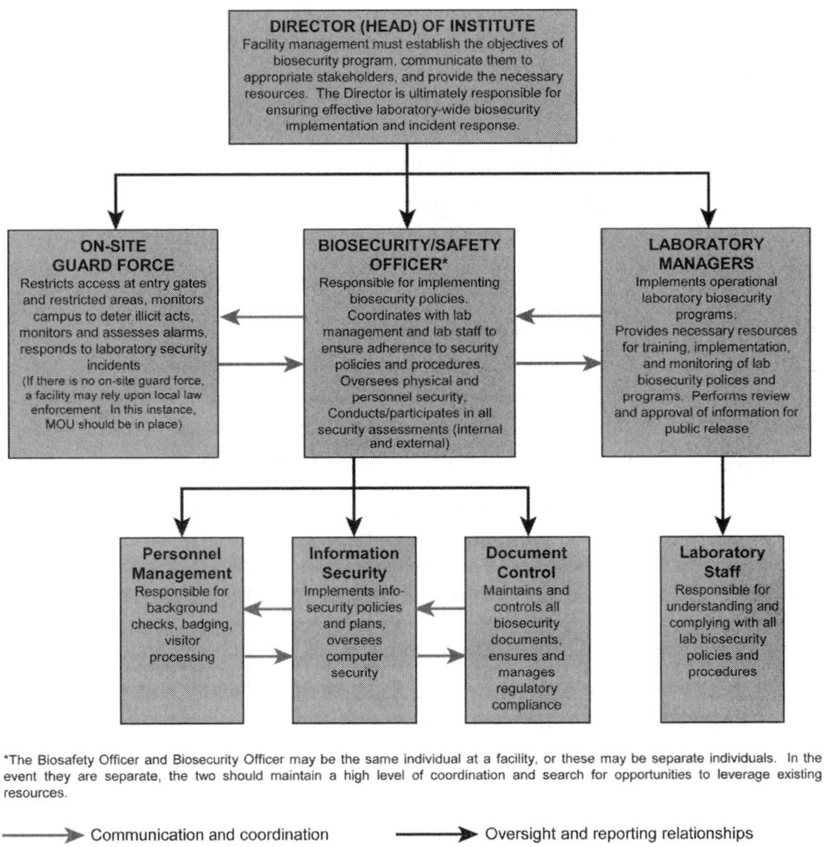

FIGURE 4.2 Program management roles and responsibilities.

4.2 BIOSECURITY SYSTEM DESIGN

Management must be responsible for knitting together all of the components of laboratory biosecurity into a comprehensive and appropriate biosecurity program. To achieve this, the design of the biosecurity system is necessarily based on the risk assessment. Before any documents are created, managers must work together to establish clear and succinct program objectives. Specifically, managers must identify which scenarios to protect against (the unacceptable risks) and which scenarios to be prepared to respond to (the acceptable risks). This differentiation is essential, given the challenges associated with securing dangerous biological agents. Managers should also consider whether the biosecurity system will be designed to accomplish any auxiliary tasks, such as the protection of computers and other expensive equipment.

For most biological science facilities, the most appropriate objectives of the biosecurity system may be to deter the authorized Insider and deny unauthorized Insiders and low-level Outsiders the ability to execute those scenarios whose risks are assessed to be at the moderate level or above. Denial or containment of a moderately equipped Outsider adversary is too resource-intensive for most biological institutions to support, even for those assets that may be considered high risk, so deterrence of these adversaries may be the only achievable course.

As demonstrated in Chapter 5, "Specific Biosecurity Recommendations," biosecurity should always be implemented in a graded manner. This helps ensure effective use of resources and helps prevent unnecessary impact on the bioscience institution's mission. A graded protection system is achieved by increasing security incrementally and forming concentric layers of protection around the facility's assets based on the level of risk associated with these assets (Figure 4.3). The layer within which an asset resides should correspond to the level of security it requires. The layers should have increasing physical, personnel, and MC&A requirements; details on which controls may be appropriate for which levels are in the individual component sections. Information and transportation security requirements will also vary based on the risk assessment, but these controls are

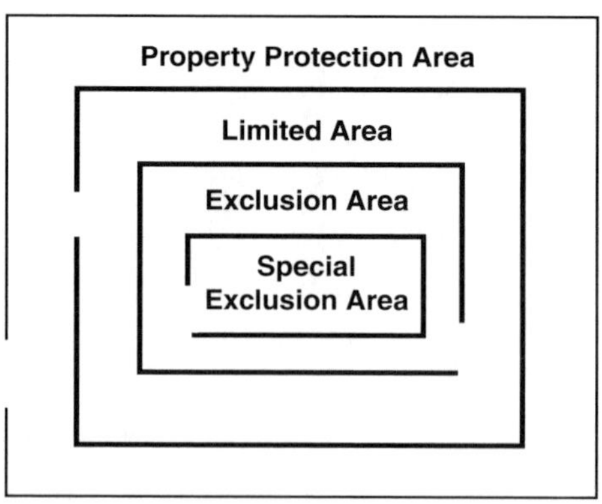

FIGURE 4.3 Graded protection areas.

not necessarily specific to physical areas within a facility. From the lowest security requirements to the highest, the nested levels of protection are referred to as *Property Protection Areas*, *Limited Areas*, *Exclusion Areas*, and *Special Exclusion Areas*. Limited, Exclusion, and Special Exclusion areas are always considered to be restricted areas as the term is used in the following sections.

Property Protection Areas are the outermost layer and are appropriate locations for low- and very-low-risk assets. A Property Protection Area is defined by the outermost perimeter of a facility's campus or by the facility walls if there is no discernible exterior perimeter that belongs to the facility. For example, a Property Protection Area may contain buildings that require little or no protection measures, such as warehouses, some administrative offices, and public access areas. low malicious use risk agents may be used and stored in Property Protection Areas.

A Limited Area resides within a Property Protection Area or within another Limited Area. There may also be multiple, separate Limited Areas contained within a given Property Protection Area. A Limited Area is contained within walls, doors, windows, or other barriers that identify its boundary. An entire building may be designated as a Limited Area, or individual rooms or laboratories located within a building may be designated as Limited Areas. They are appropriate locations for moderate-risk assets. For example, a Limited Area may be suitable for the storage and handling of moderate malicious use risk agents and sensitive information associated with biosecurity. Access to a Limited Area should require access authorization and a unique item, such as a physical or electronic key, or accompaniment by an authorized escort. Electronic access control systems are preferred for Limited Areas. Those areas without electronic intrusion detection systems should be monitored periodically by roving personnel who are specifically checking for security issues, e.g., signs of unauthorized access or unauthorized personnel. A visitor to a Limited Area should remain under escort unless the visitor has met the requirements for unescorted access. Couriers and other delivery personnel should either deliver their packages to an area where they do not require an escort, leave the delivery with a guard, hold authorized access, or be escorted.

An Exclusion Area should reside within a Limited Area or within another Exclusion Area. There may be multiple, separate Exclusion Areas within a Limited Area. An Exclusion Area, like a Limited Area, is contained within walls, doors, windows, or other barriers that identify its boundary. Exclusion Areas are appropriate for the storage and handling of high malicious use risk agents (including HMUR-infected animals), the central alarm monitoring station, and sensitive biosecurity information. An Exclusion Area should have access controls and intrusion detection to provide reasonable assurance that only authorized personnel are allowed to enter and exit the area without escort. Access to an Exclusion Area should require access authorization and a unique item and unique knowledge, such as a physical key and positive identification provided by a guard or an electronic key and a PIN. The keys to these areas should be controlled, and those individuals in possession of a key should be documented (Figure 4.4).

An Exclusion Area generally has a smaller set of individuals who are authorized to enter it than in the case of a Limited Area. Individuals who have a legitimate

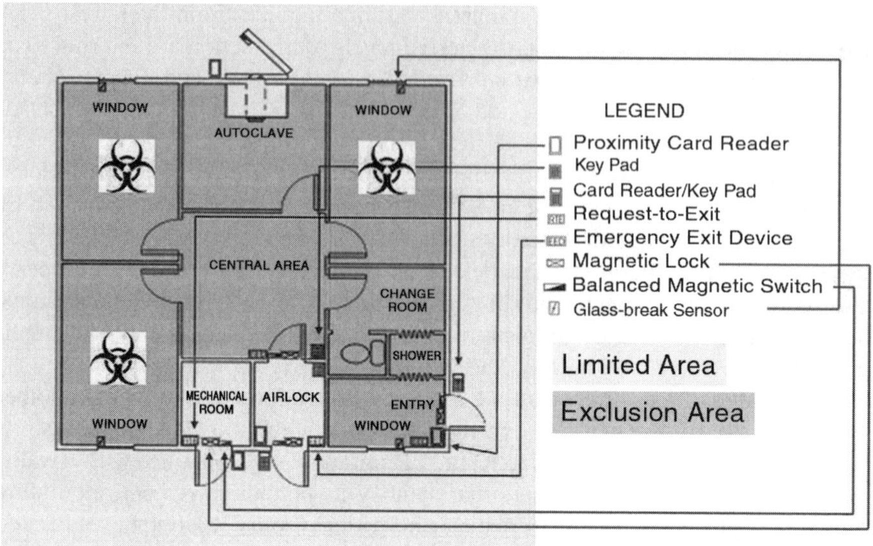

FIGURE 4.4 **(See color figure following page 16.)** Example of a physical security layout.

purpose for entering an Exclusion Area but who do not have routine access privileges should be accompanied by an authorized escort. Authorized personnel, visitors, and escorts should be required to sign an entry-and-exit log if electronic logging is not provided by the access control system. If electronic logging is provided for authorized personnel but not for visitors, a visitor log should be used to capture the date and time of the visit, the visitor's name, organizational affiliation, purpose of the visit, badge number (if applicable), and the escort's name.

Those individuals who are not employed by the institution but who have been invited to work in an Exclusion Area should meet the same criteria as employees of the institution who have unescorted access to it. These criteria should preferably be met prior to the individual's commencement of work in an Exclusion Area, but if this is not possible, the individual should remain under escort until such time as unescorted access may be granted. Casual visits to Exclusion Areas should be discouraged; however, guided tours and other demonstration-related activities may be occasionally necessary. Under these circumstances, great care should be taken to secure all dangerous pathogen sources and to keep each visitor under escort at all times.

Storage containers, such as freezers or refrigerators, may also be considered Exclusion Areas when located within a Limited Area and access-controlled with both a unique item and the unique knowledge of authorized individuals. Storage containers may also be contained within Exclusion Areas and access-controlled with both a unique item and unique knowledge to further reduce the number of personnel with access to the materials inside so as to distinguish them from those with authorized access to the larger Exclusion Area. In this case, both the storage container and the area in which it resides would be considered Exclusion Areas.

For those facilities maintaining Exclusion Areas, security personnel should monitor all entrances and exits either in person or through the use of an electronic intrusion detection system, which communicates to a centralized alarm communication and display system.

A Special Exclusion Area should reside within an Exclusion Area. It is contained within walls, doors, windows, or other barriers that identify its boundary. Special Exclusion Areas are appropriate locations for extreme malicious use risk agents. It is anticipated that the majority of bioscience facilities do not hold extreme malicious use risk agents, so the discussion of Special Exclusion Areas here is limited. The Special Exclusion Area envelope should be reinforced and strengthened to impede an adversary attempting to breach this envelope. It should have electronic access controls and intrusion detection systems to provide access to only authorized individuals. Access to a Special Exclusion Area should require access authorization and a unique item and unique knowledge. Biometric verification should be strongly considered for the access control system. No personal or casual visitors should be permitted in Special Exclusion Areas without special permission, escorting, and securing or removing EMURs. If maintenance of the Special Exclusion Area is required, all EMURs should be secured or removed in advance, and maintenance personnel should be escorted at all times. Entry of delivery and custodial personnel should be prohibited. Working visitors should be screened to the level of authorized employees, and like regular staff, they should not be permitted access until the authorization process has been successfully completed. Working visitors should be subject to a preliminary screening equivalent to a prehire staff screening and, like regular staff, should not be permitted to work alone in Special Exclusion Areas.

Consolidation of dangerous pathogens and toxins, to the extent possible, facilitates cost-effective implementation of biosecurity. Such consolidation provides a means for limiting the number of restricted areas and associated biosecurity protocols.

4.3 RESPONSE FORCE

The response force should perform its duties according to a prearranged response plan. Establishing a memorandum of understanding or a specific agreement between the facility and the local law enforcement (LE) agencies significantly enhances cooperation. This agreement should define the conditions under which a response force will be summoned and the protocols to follow thereafter. An example of a MOU is provided in Appendix D.

Determining whether a facility needs to have an on-site guard force should be based on the risk assessment. The role of the guard force should be to provide assessment and response services for any security incident that may occur at the site. Other duties may include monitoring electronic access control and intrusion detection system alarms, patrolling the property, verifying the identity of authorized individuals and their vehicles, and managing deliveries and visitor traffic. The on-site guard force should have clear guidelines that articulate when local LE should be summoned, under which conditions force may be used, which areas contain hazardous materials, and the protocols for addressing security incidents in these areas.

Local LE may be the police or other local, regional, or national security forces that are trained to manage a security situation involving dangerous materials. Facilities that handle dangerous pathogens and toxins should ensure that all emergency response personnel, including local LE, are aware of their unique safety issues and what protocols to follow if an incident occurs.

If there is no on-site guard force, and local LE is to be relied upon for security incident response, an MOU should be drawn up that describes the circumstances under which LE personnel may be summoned, the protocol to follow once on site, and the scope of authority for all parties involved. Response times should be appropriate for the protection strategy employed at the site. If, for instance, the facility holds high malicious use risk agents, and local LE cannot respond to an alarm on site within a reasonable period of time, the facility might consider employing an on-site guard force. The response force should be equipped and authorized to confront an adversary. Local LE can reinforce the on-site guard force if one is present; under these circumstances, chain-of-command issues are particularly important to establish in advance.

Local LE should be introduced to those responsible for site security and dangerous pathogens, provided with an orientation of the facility, and should attend site security awareness training and participate in joint exercises with facility staff to rehearse incident response plans.

To be effective, response force personnel should have training on their equipment, use-of-force procedures, emergency response procedures, and the biosecurity system. They need to be familiar with the features of the facility and its operations, understand restricted area access, have specific instructions on the limits of their authority, and be aware of pertinent biosafety issues. Responders may also need to be subject to personnel screening measures. Responders should also have an up-to-date contact list of officials who require notification in the event of a security breach or hazardous situation.

4.4 PERFORMANCE TESTING

Performance testing allows the effectiveness of the entire physical security system — equipment, policies, procedures, and people — to be evaluated. It is analogous in many ways to commissioning a containment laboratory. Performance testing of the physical security system should be conducted on a regular basis to ensure that the system is generating reliably and effectively. A performance test should be based on a plan that incorporates all of the policies, procedures, and hardware components of the physical security system. The performance test plan should specify each element of the system and what tests must be conducted in order to ensure that the system is performing as intended. The plan should include testing of integrated systems of equipment and hardware, administrative procedures, and protective force procedures for both on-site guards and local LE. Any change in the design of the physical security system or its procedures will have a direct impact on the performance test plan, which should be updated accordingly.

The frequency of performance testing may vary depending on the type of physical security system. Performance testing of a system based largely on manpower and mechanical systems, for instance, will not need to be as frequent as for electronic physical security systems. In the former, the frequency may be driven by turnover in personnel. In the latter, the number of false alarms or routine maintenance schedules should determine how often the system is tested. Regardless of the type of physical security system, a comprehensive review of all physical and procedural controls should be conducted annually. When the physical system relies upon an electronic system, performance testing should also occur annually in order to demonstrate overall facility physical security system effectiveness. Appendix E has an example of SOPs for testing the performance of electronic access control systems.

Once the performance tests are complete, the results should be documented and corrective measures should be taken for any elements that are not functioning as intended. Corrective measures may include replacing faulty equipment, training or retraining personnel, and amending existing policies or procedures.

4.5 DOCUMENTATION

It is important for management to develop various official laboratory biosecurity documents, including a biosecurity risk assessment, policy statements, a biosecurity plan, and SOPs. The goal of these documents is to demonstrate institutional support, document the chain-of-command and responsibilities, clarify consequences for noncompliance, and ensure institutional memory.

The risk assessment should be a written document; this document may be either a stand-alone report or incorporated into the laboratory biosecurity plan. A formal site-specific risk assessment should be conducted regularly and not less than once every 5 years. If the operations or materials at the facility change substantively, or information from LE or intelligence agencies suggests modification to the threat potential, the security systems should be reevaluated for effectiveness against the revised risk. A facility official who is responsible for laboratory biosecurity should review the risk assessment annually (or as circumstances change).

The laboratory biosecurity plan should document all elements of the laboratory biosecurity system — from its objectives to its implementation. It is analogous to the laboratory biosafety plan, which should also exist at every bioscience institution that works with dangerous pathogens or toxins. There may also be other important plans, or manuals, at these institutions that address subjects such as animal care and waste disposal.

At every bioscience facility that works with dangerous pathogens and toxins, SOPs should exist in addition to the biosecurity plan. SOPs are detailed documents that specify detailed procedures or operations that may only be referred to generally in the biosecurity plan. For instance, biosecurity procedures that might warrant an SOP include laboratory housekeeping, equipment maintenance, spill cleanup, sensitive information handling, guard-force post orders and response protocols, and pathogen transport procedures.

At large or complex facilities, it may also be important to have high-level management policy statements that clarify the objectives of the biosecurity system, the requirements of the specific components of the biosecurity system, and management's role and responsibility for biosecurity oversight.

4.5.1 Laboratory Biosecurity Plan

Every facility that stores, uses, or transports dangerous biological agents should develop and implement a laboratory biosecurity plan that establishes systems, policies, and procedures to ensure the security of areas that contain dangerous biological agents. Appendix C has a template for a biosecurity plan.

A laboratory biosecurity plan should describe in detail all the objectives, strategies, elements, and procedures associated with the facility's laboratory biosecurity system. It should be based on a facility-specific risk assessment. In other words, it should articulate which biosecurity risks the facility management finds unacceptable. The plan should then document how the facility protects against these unacceptable biosecurity risks. The laboratory biosecurity plan may also explain which biosecurity risks management finds acceptable. It should indicate that incident response plans are in place to defend against these acceptable biosecurity risks. The explanation of the unacceptable and acceptable risks in the plan serves to establish the specific objectives of the biosecurity system.

In addition to documenting the risk assessment, or at least referring to its conclusions, the biosecurity plan should describe the following elements of the laboratory biosecurity program: management responsibilities, physical security, personnel management, MC&A, transport security, and information security. All procedures related to biosecurity and the consequences for failing to follow these procedures should be established and documented in the facility's biosecurity plan.

Specifically, the plan should describe, justify, and document the graded protection provided to dangerous biological agents and indicate how the system will detect, deter, and respond to unauthorized access to these agents. The plan should provide requirements for notification (both oral and written reports) in the event of the theft, loss, or release of dangerous biological agents. It should have provisions for routine cleaning, maintenance, and repairs of restricted areas.

A facility may choose to prohibit certain items on the property. If so, this should also be documented in the laboratory biosecurity plan. These items may include weapons, explosives, and other dangerous instruments or material likely to produce substantial injury or damage to property; alcohol; controlled substances, such as illegal drugs; other items prohibited by law; and electronic recording devices.

The document should reference the laboratory biosafety manual and relevant incident response plans. This document should provide an explanation of the required laboratory biosecurity training to be received by all personnel. Personnel should be trained to follow those portions of the laboratory biosecurity plan that apply to their professional responsibilities. It may not be necessary to train all personnel on all aspects of the plan.

The biosecurity plan should be site specific. For institutions with multiple sites, there may need to be both an institutional and a site-specific biosecurity plan.

The laboratory biosecurity plan should be reviewed and performance-tested annually, and revised if necessary. This process should be repeated if a security incident occurs or if any of the assumptions of the biosecurity risk assessment change.

4.5.2 Security Incident Response Plans

Security incident response plans should address at least three types of issues:

1. Routine security system management of security incidents
2. Reports by LE or the intelligence community that indicates that the laboratory is or may be targeted by a particular adversary
3. Those scenarios determined to represent acceptable risks and thus not specifically addressed by the security system, especially those that may have high consequences but have a low threat potential

Routine security system management requires the facility to have procedures in place for responding to alarms, indications of forced entry, and suspicious activity. The security incident response plan should describe protocols for each of these situations. For each situation, the plan should:

- Clarify who is responsible for responding to and assessing alarms generated by the security system and the approved communication channels for reporting the results of that assessment.
- Identify the responsibilities of personnel who would report and receive notification of an incident. Personnel should report the incident through the appropriate communication channels to the designated individual or office.
- Identify the responsibilities of the response force. The plan should address communication needed among the force and the method the response force will use to respond. It is important that the plan describe the circumstances under which the response force is authorized to use force, and whether the response force has the authority to arrest unauthorized individuals. If the response force does not possess arrest authority, the plan should specify how the response force will detain authorized individuals until local LE officers arrive.

- Identify specific personnel who are responsible for alerting local LE, and the circumstances that require making this notification.
- Identify personnel responsible for interacting with local LE officials upon their arrival.
- List up-to-date points of contact to be used for incidents that occur after normal business hours. This contact list should be checked for accuracy on a regular basis to ensure that contact information has not changed.
- Clarify unambiguously the personnel who are authorized to bear arms and the circumstances in which the arms may be drawn and used.
- Provide further clarification on additional procedures to be followed in specific types of security incidents.

Whenever the security system generates an alarm, it must be assessed for legitimacy. The security incident response plans should clarify personnel responsible for alarm assessment. The plan should also describe whether the central alarm monitoring station — if it exists at a facility — will be manned during alarm assessment. If only one member of the guard force works at the guard station at a time, and he or she is responsible for leaving the station to assess an alarm, procedures should be established for the reporting of additional alarms while the guard is absent from the station. The plan should describe how false alarms will be identified, and the types of measures that will be taken to prevent their future occurrence. The plan should detail how all alarms and incidents will be documented.

Whenever local LE is involved, it is important to have some form of MOU in place to establish the framework for coordination. The MOU should address response times and priorities for a variety of situations, including incidents that are not associated with dangerous pathogens (e.g., vandalism, petty theft) and those incidents that may indicate theft or attempted theft of the facility's dangerous pathogens. It should detail the procedures for arrest and/or removal of individuals from the premises. LE needs to be informed about the issues surrounding biological laboratories and have a point of contact at the laboratory that will assist in the response to any incident in or around a containment space. The MOU is the ideal way to document this information. The issue of information sharing can also be addressed in an MOU. Will the facility provide local LE with details regarding the location of its most dangerous pathogens and who has access to them? If so, how will local LE protect that information?

4.6 ASSESSMENTS AND AUDITS

Once a biosecurity system has been designed and put into practice, management must continue to ensure its successful operation. This can be accomplished through a self-assessment process. Self-assessments are internal audits that provide facilities with internal monitoring of laboratory biosecurity programs to ensure ongoing compliance with laboratory biosecurity requirements. A self-assessment should cover the following laboratory biosecurity elements: management responsibilities, physical security, personnel management, MC&A, transport operations, and information security. Any laboratory biosecurity records, such as laboratory notebooks, inventories, and databases, should be available for review at any time by personnel who are responsible for conducting self-assessments.

Depending on a country's laws or regulations, facilities may also be required to submit to audits by an external agency. Self-assessment and external audit requirements should be documented in the laboratory biosecurity plan and, ideally, the self-assessment requirements should be more onerous and frequent than the external requirements.

The program management self-assessment should ensure that laboratory biosecurity, emergency, and incident response plans are regularly reviewed and revised appropriately. It should also ensure that all staff numbers are compliant with their training requirements. In addition, this self-assessment should ensure that restricted information is secured appropriately and that records are accurate and up to date.

The physical security self-assessment should verify that all security devices are operational and that maintenance is actually performed on a regular basis. Performance testing of the physical security system (Section 4.4) is one element of a physical security self-assessment. This should also include an assessment of response forces to ensure training is conducted regularly, that all equipment functions properly, and that the force is prepared to respond.

The personnel management self-assessment should ensure that granting of personnel access to facilities is appropriate and up to date. This self-assessment should address these questions: Have all individuals with access undergone the appropriate personnel screening? Are there any individuals that have access but no longer require it? Personnel access applies to all employees and visitors. The self-assessment should address the components of the facility's personnel management program, which may include key elements such as access authorization, visitors, badges, hosting, escorting, and travel. The self-assessment should also include a review of all access logs, both electronic and handwritten.

The MC&A self-assessment should review all aspects of the MC&A program. This assessment will include taking a physical inventory and comparing it to the inventory records. Methods used during physical review or during inventory reconciliation of HMURs should include counts of the entire inventory or, for lower-risk materials, a statistical sampling of records and repository materials. How are materials outside of the repository accounted for? Are accountable individuals assigned to all appropriate materials? Do the accountable individuals properly execute their responsibilities? Review of inventory records should be conducted at least annually and, if stipulated in the biosecurity plan, may be conducted randomly or without prior announcement.

The transport security self-assessment should include a review of transport SOPs, chain-of-custody documents, and other shipping records. This is also the time to verify that transport practices are carried out in accordance with international and national standards, SOPs, and other relevant documents.

The information security self-assessment should ensure that effective security policies and procedures are being used to protect sensitive information at a facility. The assessment should evaluate the ways in which sensitive information is identified, handled, stored, and transmitted at a facility, as well as the policies and procedures used to govern these activities. This evaluation should encompass a review of information security for both paper and electronic media.

All self-assessments and external audits should be documented in a written report. Whenever a report contains one or more findings of error or ineffectiveness, corrective actions should be undertaken and documented. The corrective action plans should include specific changes to be made, organizations and individuals responsible for each change, a process for tracking key milestones, and steps to verify the effectiveness of actions prior to closure. It is essential that upper management is involved in corrective actions in order to resolve findings and thereby improve the effectiveness and efficiency of the laboratory biosecurity program. Once the corrective action has been taken and tests for effectiveness have been successfully conducted, a record of closure should be created.

Oversight of laboratory biosecurity is not limited to self-assessments and external audits. Reporting by staff and laboratory managers also helps ensure that the laboratory biosecurity system is operating effectively. Although routine reports incur extra administrative overhead, which limits their frequency and detail, they nevertheless help demonstrate good laboratory biosecurity practice. Continuing attention is being paid to the practice of good laboratory biosecurity, and may be required for compliance with external regulations.

Anything out of the ordinary should not rely on routine reporting, however. In that case, a mechanism for nonroutine reporting needs to be created to ensure a timely response. It is essential to foster an environment that encourages anyone to report anomalies, such as unjustified access to a restricted area.

4.7 TRAINING AND EXERCISES

Training and exercises are both crucial for the effective implementation of laboratory biosecurity. Both provide the opportunity to familiarize employees with the operation of laboratory biosecurity, reducing the possibility of unsecure practices. Because laboratory biosecurity relies predominantly on specific practices by laboratory personnel — rather than sophisticated technical equipment — it is especially important to review these protocols and procedures on a regular basis.

Training sessions, of which there are several types, acquaint laboratory personnel with proper procedures and reporting requirements by highlighting critical elements of the laboratory biosecurity plan, policies, and any other laboratory biosecurity-related directives. These sessions should also clarify where employees can find further guidance on laboratory biosecurity. Similarly, exercises provide an opportunity for the guard force and local LE to practice responding to a laboratory biosecurity incident, ensuring that these groups can cooperate successfully in the event of an actual incident.

All employees at the facility should be trained on the basics of laboratory biosecurity. Each individual who works in or visits areas where dangerous biological agents or restricted information are stored or handled should receive specific laboratory biosecurity training. The training should be tailored to an individual's roles and responsibilities. Function-specific training is especially important for staff that has access to MMURs or HMURs. Supervisors and the response force also need specialized training. All of this general and specialized training should occur, at a minimum, upon hiring and whenever there are new circumstances, such as a change in duties and/or access authorizations. Ideally, security refresher training should be conducted annually. Failing to follow established security procedures should result in remedial training, disciplinary action, termination, and/or criminal prosecution, as appropriate.

4.7.1 Specific Training Topics on Laboratory Biosecurity

All applicable aspects of the laboratory biosecurity system should be included in this annual training. All employees should be informed that there are certain types of restricted access areas at the facility — Property Protection, Limited, and Exclusion Areas — and the applicable access control procedures for each area. It may or may not be appropriate to reveal the location of all Limited and Exclusion Areas to all employees. Every employee should be informed of general physical security policies and procedures at the facility regarding access hours, visitor logs, vehicle security, tailgating prohibition, suspicious activity and unauthorized individual reporting, and prohibited articles. Employees should be informed of the following personnel security measures at facilities: position designations, personnel screening requirements, derogatory information that could affect access, badges, visitors, host responsibilities, escorting, and foreign travel requirements. Employees need to understand information security protocols, including types of restricted information, restricted information access and protection requirements, network protections, user desktop system protections, remote access, and wireless networking.

In come cases, particularly at laboratories with HMURs or EMURs, employees and long-term working visitors should receive general counterintelligence awareness training once a year. The goal of this training is to assist staff in identifying situations or actions by persons outside the laboratory who may intend to obtain information or materials from within the laboratory without authorization. Any individual who attempts to gain information about the facility's operations, biosecurity measures, capabilities, or personnel, who does not have a legitimate need for this information, should be reported. Solicitation attempts of this nature may be made by mail, e-mail, fax, telephone, or in person.

All employees should be informed of the policy regarding incidents and violations and the associated reporting procedures. Employees should also be trained on how to handle media requests. Is there an institutional point of contact to which they should refer the person making the inquiry, or are they authorized to speak on behalf of the institution on certain topics?

Personnel need to be trained on their responsibilities associated with any unique credentials that they control. They must understand that any items that provide an entrance to a restricted area, such as a badge or proximity card, should remain in the possession or control of the individual to whom it has been assigned. The credential should not be shared with anyone, and loss or theft of the credential should be reported immediately. The credential should be returned to the institution upon termination or transfer. Unique information, such as a PIN, should never be shared. Individuals who are found to have knowingly violated these procedures should be cited for a security infraction and retrained or disciplined as appropriate.

Entering a restricted area without a proper badge and access authorization should constitute a security infraction. Anyone with access authorization is responsible for ensuring that individuals who have lost access authorization or who are unauthorized are not permitted entry into restricted areas. Unauthorized individuals may be recognized by the absence of a badge or simply by unfamiliarity. An individual who forgets a badge is responsible for obtaining a replacement prior to accessing restricted areas. Managers should be attentive to whether an individual repeatedly requires badge replacement and, if so, address the matter as an issue of security concern. It is important to instill in the workforce a sense of security responsibility associated with badges used in facilities with high or extreme malicious use risk agents. Employees should also be counseled not to wear their badges in public, use them for unofficial identification, or leave them in an uncontrolled environment. Badges should be treated as valuable items and protected accordingly.

Tailgating, defined as more than one person passing through a controlled access point on a single credential, should be prohibited. Although this may be a difficult provision to implement, it is important. Often individuals think it is a courtesy to hold a door open for another individual or, if someone knocks, to simply let the person enter. The perception is often that it is unreasonable to insist that an individual present personal credentials, especially if the individual desiring entrance is familiar. What must be explained to personnel is that an authorized individual can never be certain that another person is still in possession of current access authorization. This can only be determined by visually inspecting that person's security credentials. In order for these policies to be fully effective, security credentials must be actively managed; keys must be returned upon termination or transfer, and electronic access must be terminated or modified as access requirements change. The term *tailgating* does not refer to granting access to an authorized visitor under escort. In this case, the escort is acting as the security credential for the visitor.

A similar circumstance may occur if a facility determines it is acceptable to permit the act of "vouching." Vouching is not a recommended practice but, if it is permitted, an authorized individual may "vouch" for the access authorization of another and permit that person to enter a restricted area. As mentioned previously, however, the individual who vouches for another cannot be certain of the state of the individual's access authorization and, if an error occurs, the one who vouched for the credentials of another should be held responsible for any security violations that may ensue. Rather than permit the act of vouching, facilities should require those who have lost or forgotten their security credential to obtain a temporary pass. This temporary pass may or may not provide an individual access to an area (e.g., it may not be electronically encoded as a key), but it will provide those who have authorized access a valid premise for allowing the individual to enter the area. Once inside a restricted area, if electronic logging of the temporary pass information is not possible, the individual should be required to log in as a visitor.

Personnel who work with dangerous pathogens and toxins should be trained to secure those agents in the event of an emergency if they believe that they will not put themselves in danger of injury. If the situation presents imminent danger, personnel should evacuate and inform security officials of the situation once it is safe to do so. Security officials should be told which materials were left unsecured and where the unsecured materials are located. Procedures to follow in the event of an emergency should be provided to both facility personnel and emergency responders.

Security personnel should be immediately notified if an individual is identified inside a restricted area without proper authorization. This situation could occur when an individual is noticed either without an identification badge, without the proper authorization level indicated on an identification badge, or as unfamiliar to the regular staff. If the situation appears threatening, only security personnel should approach the unknown individual. If the situation appears nonthreatening, it may be appropriate for the regular staff to ask who the person's escort or host is, or whether the person is lost. If an unauthorized individual has been left unattended by an escort or is lost, an authorized individual should escort that person out of the restricted area and either remain as the individual's escort or turn the person over to security personnel, an appropriate manager, host, or another escort, until the situation can be satisfactorily resolved. An escort who leaves a visitor in a restricted area unattended should be cited with a security violation and, at a minimum, should receive remedial training on proper security procedures.

If possible, it is important to determine how the unauthorized individual entered the restricted area without an escort prior to releasing the individual from official supervision. Establishing the protocols for a broad range of circumstances that may arise when an unauthorized entry occurs is helpful for those personnel who are authorized to be in restricted areas and for those who are responsible for responding to such incidents.

The probability that an unauthorized individual will be able to gain undetected entry into a restricted area will be influenced by the level of security training that facility personnel receive and their willingness to support the established security protocols. Personnel should be trained to report suspicious activity to security personnel. Such activity may include an individual displaying strange behavior or being suspiciously out of place; an individual observing or photographing the facility site or operations, annotating maps, or using binoculars; or a vehicle parked or operated in a suspicious manner on, or in the vicinity of, the facility.

4.7.2 Training for Supervisors

Supervisory training should be required annually for the managers at facilities. The training should inform managers of their reporting responsibilities and should encourage them to report any information that raises doubts about an employee's continued eligibility for access to dangerous biological agents or restricted information.

Supervisors should also be trained to monitor the physical and mental health of those who work with dangerous biological agents. If an employee appears to be in poor health, it is possible that the employee may pose a safety or even a security threat. That person should be counseled and prevented from working with dangerous biological agents until fully recovered.

Supervisors should be trained to know what forms of employee assistance programs are available to employees and be trained to spot abnormal behaviors that may indicate a need for those types of services.

Supervisors should be trained to manage those items owned by the facility that are issued to employees. It is especially important for supervisors to manage items that are directly related to access control or other security functions, such as a badge.

Supervisors should be trained to manage any aspect of personnel screening for which they are responsible, to maintain training and immunization records, and to review and approve information for public release. Some of these functions may be the responsibility of other staff, such as human resources. If so, supervisors should ensure the compliance of their employees in meeting these requirements.

4.7.3 Response Force Training

If a facility has an on-site response force, it should receive task-specific training regularly, or at least annually. The objective of the response force training and qualification program should be to develop and maintain the required competencies of those personnel.

Prior to receiving an initial assignment, an on-site response force officer should successfully complete basic training. Facility-specific training requirements should include the location of, and security and safety issues associated with, any dangerous biological agents at the facility as well as the individuals who are accountable for those materials. The response force training could also include, but not be limited to, the following types of instruction: orientation and standards of conduct; laboratory biosecurity education and operations; dangerous biological agent protection requirements and issues surrounding access to restricted areas during an emergency; restricted information protection requirements; response to and reporting of incidents of laboratory biosecurity concern; protection of government property; physical fitness training; facility operations familiarity; safety training; legal requirements and responsibilities; weaponless self-defense; use-of-force policy; communications, including methods and procedures; vehicle operations, including methods and procedures; post and patrol operations; and use of assigned personal protective equipment (PPE).

Exercises of various types should be included in the training process. The types and frequency of training exercises should be approved by the facility laboratory biosecurity officer. At a minimum, the following elements should be included in the training exercise program:

- Exercises that involve each response force officer's shift should be conducted monthly. These exercises should be planned and conducted so as to provide facility-specific training.
- Exercises with local LE or other personnel who would assist the on-site response force during a facility laboratory biosecurity incident should be conducted at least annually.

Reports of each exercise summarize results and problem areas, should be prepared for management review to support response force training, planning, and corrective actions.

5 Specific Biosecurity Recommendations

This chapter describes specific biosecurity recommendations for facilities that have different levels of risk. Laboratory biosecurity risk mitigation measures, as described in Chapters 3 and 4, should be more robust for higher-risk scenarios than for lower-risk scenarios. In this chapter, facilities are characterized by their highest-risk biosecurity scenarios. For example, we define a "Moderate-Risk Facility" (MRF) as one whose risk assessment determines that the facility only has low- and moderate-risk biosecurity scenarios to protect against. Most bioscience institutions will likely be characterized as low- or moderate-risk facilities.

These recommendations are based on the authors' experience implementing laboratory biosecurity and presume a strategy of deterrence, not denial or containment. The risk mitigation measures are progressive and supplementary: Recommendations for each level presume that those recommendations for the prior level are already in place. A chart summarizing these recommendations appears in **Tables 2-5**).

5.1 LOW-RISK FACILITY

Low-risk facilities (LRF) have low-risk biosecurity scenarios as their highest-risk scenarios. If properly implemented, laboratory biosafety measures coupled with best business practices often provide sufficient security at this level.

5.1.1 Program Management

The Principal Investigator should be responsible for ensuring that the laboratory has appropriate biosecurity. He or she should ensure that all personnel working with low malicious use risk agents are aware of the necessary laboratory biosecurity and biosafety measures.

5.1.2 Physical Security

LMURs may be used and stored in a Property Protection Area and should be protected to the degree provided by good biosafety practices. For example, access to the laboratory or storage area should be restricted in some way (e.g., doors are locked), especially when an authorized individual is not present.

5.1.3 Personnel Security

Personnel authorized to work with or handle LMURs would be considered low-risk positions. As such, institutions may feel comfortable employing these individuals with minimal background checks. It is good business practice to verify the employment history and educational background of job candidates and to have a human resource system in place that can respond to abrupt changes in an employee's behavior at work. It is prudent from a safety perspective to ensure that visitors (i.e., personnel who do not work in the laboratory) are allowed into the laboratory only with permission of the PI or under escort of an authorized person.

5.1.4 Material Control and Accountability

Standard laboratory records, such as laboratory notebooks, should be used to document the storage and use of LMURs. A laboratory notebook maintained by the accountable scientist should document the stocks and use of materials under his or her purview. These measures are simply good laboratory practice.

5.1.5 Transport Security

The PI should be aware of all internal or external transport of LMURs. The transport of LMURs should also be documented. This could be a record in the laboratory notebook. Recorded information should specify what was shipped, the shipment date, and the recipient.

5.1.6 Information Security

Prudent policies should be in place for computer network security, passwords, and e-mail use.

5.2 MODERATE-RISK FACILITY

Moderate-risk facilities (MRFs) have moderate-risk biosecurity scenarios as their highest-risk scenarios. MRFs should implement the measures for LRFs plus the elements described in the following text. Here, the concept of graded protection should be incorporated because almost all MRFs will have both low-risk and moderate-risk biosecurity scenarios, and LMURs generally do not warrant the same level of biosecurity as moderate malicious use risk agents.

5.2.1 Program Management

Policies and procedures for biosecurity should be established that address all components: physical security, personnel security, MC&A, transport security, information security, and program management. Adequate funding should be allocated for biosecurity implementation, training, and review. Risk assessments should be conducted, and security policies, procedures, and plans should be documented and reviewed annually. Security incident response plans should be written, implemented, and exercised. Corrective measures should be taken when reviews show deficiencies.

There should be a designated biosecurity officer that oversees the implementation of appropriate policies and procedures. For small facilities with limited MMUR programs, program management roles can be assigned to a member of the research team. The biosecurity officer should ensure that all authorized users of MMURs have appropriate biosecurity training. A facility representative should conduct self-audits of the biosecurity program to ensure compliance. Penalties and/or remedial education should be established and enforced for breaches of the program.

Management should be attentive to personal issues that may have security or safety implications, and it should provide counsel to those individuals who have such issues or refer them to professionals who can provide them appropriate support. Managers should be empowered to deny access to individuals based upon employee mental or physical health.

5.2.2 Physical Security

For those facilities with moderate-risk biosecurity scenarios as their highest-risk scenarios, only two layers of restricted areas should be necessary: a Property Protection Area and a Limited Area. Access controls should provide reasonable assurances that only authorized personnel can enter areas where MMURs are stored or in use (e.g., controlled keys or electronic access controls).

The facility's outermost boundary should be demarcated with a fence or signage. Vehicle and pedestrian entrances should be locked during nonbusiness hours, and controlled keys should be used for access. Visitor and employee parking should be separated, and authorized vehicles should be provided with a placard, identification tag, or other identifying marker. Unauthorized personnel found in restricted areas should be identified and removed.

A Limited Area is appropriate for storage and use of MMURs. A Limited Area resides within a Property Protection Area. LMURs can be stored and used in a Property Protection Area. An entire building can be designated as a Limited Area, or individual rooms or laboratories that reside within a building can be designated as separate Limited Areas. A Limited Area should have access controls and intrusion detection in place to provide reasonable assurance that only authorized personnel enter the area. Those who are not authorized for routine access to a Limited Area should be escorted by an authorized individual and required to sign a visitor log.

Access to a Limited Area should require access authorization and a unique item, such as a physical or electronic key, or accompaniment by an authorized escort. Keys should be controlled in such a way as to ensure that they are issued only to those individuals who have a legitimate need to have unlimited access to a Limited Area. Keys should be returned to management when this access is no longer needed, such as upon transfer or termination.

Limited Areas should be monitored regularly, either electronically or by facility personnel, for signs of unauthorized access. If unauthorized access is suspected, it should be investigated (assessed). If the investigation cannot determine whether the breach was accidental or intentional or if the breach is clearly intentional, the proper authorities, including those qualified as response forces, should be summoned to investigate the situation further.

A Limited Area should have balanced strength of construction, i.e., the door should be essentially as difficult to penetrate as the adjacent wall. At a minimum, all doors and windows should be closed and locked during nonbusiness hours, and the doors and locks should be robust.

5.2.3 Personnel Security

Personnel with authorized access to MMURs should be considered to hold a moderate-risk position. Thus, conducting a basic personnel suitability check is appropriate (as well as any other investigations as required by law). This should include a prequalification check of an individual's identification and credentials prior to offering the individual a job. Once the individual has accepted a moderate-risk position, a more in-depth personnel background investigation should be conducted. We recommend that this investigation cover at least the preceding 5 years and include a drug test, personal interview, criminal check, and credit verification. There should be a process for adjudication and process for appeals. Until receiving official authorization, the individual should be escorted within Limited Areas.

Badges should be worn that indicate identity and authorization level. They may not be necessary in facilities with a small number of personnel who all know each other well, and they should not be worn under circumstances that may cause a safety concern. Visitor badges should include an expiration date.

Visitors may be allowed into the laboratory under escort of an authorized person but visitor logs should be kept. Within a Limited Area, we recommend a visitor-to-escort ratio not to exceed 8-to-1 in nonlaboratory areas, and 4-to-1 in laboratory areas.

5.2.4 Material Control and Accountability

MC&A measures for MMURs begin with good laboratory practice, such as maintaining information in laboratory notebooks, but include additional measures to address the biosecurity risk these materials represent to the facility.

An accountable individual should be identified for all MMUR stocks. Laboratory management should keep a record of where MMUR materials are stored and used, and a record of the corresponding accountable scientists.

The laboratory should use a consistent inventory methodology for all MMURs. Lab notebooks can be used to document MMUR use (who, when, what), whereas seed stock inventories should be maintained in secure, limited-access databases (electronic or hard copy). These databases should be formatted consistently (i.e., the same information is collected or organized in the same manner) throughout the facility. Inventory entries should include, at a minimum, the biological agent name/description, genus, species, strain, source, form, and responsible individual. The seed stock inventories should also include entries to reflect transfers into and out of the lab and the facility. Paper documentation of a transfer, such as carrier receipts, should be kept for historical documentation of the transfer.

Autoclaves and other forms of decontamination equipment used for MMUR materials should reside within the same Limited Area, or equivalent, as that used for laboratory and storage purposes. If this is not possible, as is often the case with incinerators, procedural control measures should be implemented. For example, it may be considered sufficient for an individual to start the decontamination process and ensure that the process will continue as intended before leaving the materials that are being decontaminated unattended outside of a Limited Area.

5.2.5 Transport Security

All transport of MMURs should be controlled and documented. Documentation of the transport in inventory records should be adequate for MMURs. Internal chain-of-custody documentation should be maintained while the MMUR is in transit within the facility. For external shipping, the sending laboratory should have knowledge of the professional capabilities of the recipient laboratory. The individuals who handle the MMUR between the laboratory and the shipping and receiving area should be screened to the same level as the laboratory personnel. Limited Area controls should be maintained on the package until transport is under way, or it should remain in the custody of an MMUR-authorized individual. A reliable and prompt shipping method should be used, and the receiving laboratory should notify the sender of successful receipt. Shipping records should be maintained for all transport of MMURs out of the facility. Procedures should be established for receipt of MMURs into the facility, including prenotification of package receipt, chain-of-custody upon receipt, and notification to the shipping laboratory upon successful receipt.

5.2.6 Information Security

Sensitive information related to the security of MMURs should be marked and handled according to documented policies and procedures. It should be protected at a level equivalent to that of a Limited Area. Potentially sensitive information should be screened prior to public release. To prevent inadvertent release of sensitive information, the means of communicating such information should be controlled (password protected or encrypted for electronic transmission, no cellular phone discussions, etc.).

Prudent policies should be in place for computer network and desktop security, use of strong passwords, and appropriate e-mail use. Staff should be trained on operational and information security measures.

5.3 HIGH-RISK FACILITY

High-risk facilities (HRF) have high-risk biosecurity scenarios as their highest-risk scenarios. HRFs should implement the measures for MRFs plus the elements described in the following text. Graded protection is critical for efficient resource allocation at HRFs because almost all of these facilities will also have low-risk and moderate-risk biosecurity scenarios, and moderate and low risk agents generally do not warrant the same level of biosecurity as high malicious use risk agents.

5.3.1 Program Management

The Biosecurity Officer should ensure that all authorized users of HMURs have appropriate biosecurity training. Basic and task-specific biosecurity training should be provided to all personnel with authorized access to Property Protection, Limited, and Exclusion Areas.

The Biosecurity Officer should conduct self-audits of the biosecurity program to ensure compliance. System performance exercises should be conducted, problems identified, and corrective actions planned and taken.

Management should be attentive to personal issues that may have security or safety implications and provide counsel to the employee or refer the employee to resources that can provide appropriate support. Managers should be empowered to deny an authorized individual access to Exclusion Areas if they believe the individual's change in behavior presents a safety or security risk. To foster a good environment for employees, there should be opportunities for job growth and appropriate compensation relative to local standards, especially for those working with HMURs.

5.3.2 Physical Security

Facilities that have high-risk biosecurity scenarios should use three layers of restricted areas: Property Protection Areas, Limited Areas, and Exclusion Areas. Exclusion Areas are appropriate for the storage and use of HMURs. Typically, personnel should pass through a Limited Area to reach an Exclusion Area.

Generally, a smaller group of individuals should be authorized to enter an Exclusion Area than a Limited Area. Access to an Exclusion Area should require a unique item and unique knowledge, such as a physical key and positive identification provided by a guard, or an electronic key and a PIN. Tailgating should be prohibited, i.e., each authorized individual must gain access using his or her authorized key, not by following behind another's authorized entry. The keys to an Exclusion Area should be controlled, and there should be documentation of each individual who has possession of a key. Storage containers, such as freezers or refrigerators that are located within a Limited Area, used only by authorized individuals and controlled by two-level access control (unique item and unique knowledge), can also be considered Exclusion Areas.

Doors into Exclusion Areas should be magnetically locked and electronically alarmed and monitored against unauthorized entry. Doors should be equipped with balanced magnetic switches and pneumatic door closers. All door hardware should be located on the interior of the door. If magnetic locks are not available, mechanical locks should be robust and keys should be controlled (i.e., keys should be only signed out to authorized individuals and returned when no longer needed).

Restricted areas should be monitored for unauthorized access. For those facilities maintaining Exclusion Areas for the storage and use of HMURs, it is recommended that security personnel monitor all entrances and exits either in person or through the use of an electronic intrusion detection system. Sensors beyond what might be employed in a Limited Area can be used to detect an attempt at unauthorized access. Such sensors might include glass break sensors on windows and motion detection or volumetric sensors within the room. Video assessment systems can be used to monitor, assess, and assist in response to intrusion detection alarms.

Individuals who have a legitimate purpose for access but who do not have routine access privileges may be accompanied by an authorized escort. Both routine and visiting personnel should be required to sign a log upon entry and exit if electronic logging is not provided by the access control system.

Doors that provide direct access to Exclusion Areas should fail securely. In other words, if an emergency alarm is activated, Exclusion Area doors should not automatically open but remain locked. Request-To-Exit (RTE) devices should be used for egress in an emergency (to preclude safety issues associated with the time required to use an authorized key). Panic hardware or other emergency exit mechanisms used on emergency doors located in an Exclusion Area should be operable only from inside the building or room, should meet all applicable life-safety codes, and have locally enunciating alarms. Emergency evacuation routes should not violate authorized access provisions, i.e., they should not go through Exclusion Areas.

The facility should have an MOU with local LE in case a response to a security incident is necessary. If the facility holds HMURs, and local LE cannot respond to an alarm on-site within a reasonable period of time, the facility should consider employing its own on-site guard force.

The Exclusion Area envelope should have balanced strength of construction, i.e., the doors should be essentially as difficult to penetrate as the adjacent walls. Windows should be locked and electronically alarmed, or reinforced (e.g., with nine-gauge stainless steel wire mesh). All other routes of entry (roof hatches, air ducts, air locks, autoclaves, etc.) should be similarly secured. All other sources of HMUR materials (e.g., autoclaves, incinerators, air filtration systems, containers, or infected animals) should be similarly controlled.

Signage should indicate what items are not permitted in Exclusion Areas (restricted items can include cameras, personal digital devices, personal cellular phones, and firearms).

Exclusion Areas should be "locked and monitored" at the end of the day. One individual should be responsible for ensuring that all HMURs are secured properly (locked); a separate individual should verify that the HMURs are indeed secure (monitored).

5.3.3 Personnel Security

Personnel with authorized access to HMURs should be considered to hold a high-risk position. Access to HMURs should be limited to those who meet personnel security standards as defined by the facility that are more restrictive than what is required for access to MMURs. It may be appropriate to conduct a prequalification check of an individual prior to offering that person a high-risk position; such a prequalification check could assess whether the individual's professional credentials are authentic, and whether the person has any record of criminal activity or terrorist or extremist affiliation. A more thorough personal background check should be conducted once the individual has accepted a high-risk position, and it should be more comprehensive than that conducted for those holding moderate-risk positions. In addition to covering all of the elements of an investigation for a moderate-risk position, we recommend that this investigation cover the preceding 10 years, include a personal interview and interviews with a broad range of associates, and explore a wider variety of issues (e.g., substance abuse, criminal check, credit verification, and anything else that might be required by specific rules or regulations). Reinvestigations should be conducted every 5 years.

Personnel at HRFs should be issued a photo identification badge that indicates the level/areas of authorized access. Badges should be worn at all times in restricted areas unless precluded by safety, in which case an equivalently strong method of authenticating access credentials (e.g., using a biometric device or PIN) should be used. Badges should not be displayed in public and should be protected in a manner similar to a valuable personal item.

Visitors should only be allowed into the laboratory under escort of an authorized person. All visitors should remain within visual and verbal range of the escort while in an Exclusion Area. If casual visitors are permitted, the escort-to-visitor ratio should not exceed 2-to-1. A working visitor-to-escort ratio should not exceed 1-to-1.

5.3.4 Material Control and Accountability

MC&A requirements for HMURs build upon those recommended for MMURs. An accountable individual should be identified for all seed and working stocks. The Biosecurity Officer or other management representative should maintain a record of the accountable individual who is associated with each HMUR agent. This documentation should track who works with what HMURs in which locations. It should also include a record of any transfers of HMUR items to other laboratories at the facility or outside of the facility.

For seed stocks, we recommend that the facility maintain inventories in secure, limited-access databases (electronic or hard copy) that are formatted consistently (i.e., the same information is collected or organized in the same manner) throughout the facility. If not included as part of the inventories, working-stock records should be maintained in laboratory notebooks. These working-stock records should include the name of the accountable individual and the names of the individual working with the agent, if different from the accountable individual; the nature of the experiment, including the status of any animals infected with HMURs; and relevant dates. Usage logs should be kept to document when the HMURs are accessed and by whom. This may be included as one element of the seed stock inventory or working-stock records.

Autoclaves and other forms of decontamination equipment used for HMUR materials should reside within the same Exclusion Area, or equivalent, as that used for laboratory and storage purposes. If this is not possible, as is often the case with incinerators, procedural control measures should be implemented. For example, an individual who is authorized to work with HMURs (or screened to an equivalent level) should remain in attendance until the decontamination process is complete and confirm that the decontamination was successful.

5.3.5 Transport Security

The level of control and documentation associated with the transport of HMURs should include all the elements of transport security for MMURs but should layer additional controls as appropriate. We recommend that the Biosecurity Officer preapprove all transport of HMURs (both internal and external). Chain-of-custody practices should be utilized for transport within a facility, including to and from shipping and receiving areas. Exclusion Area controls should be maintained on the package or it should remain in the custody of an HMUR-authorized individual if temporary storage is needed (such as in the shipping areas).

For external shipments, HMURs should be shipped by timely methods (e.g., overnight shipping). The designated individual for the sending laboratory should confirm that both the receiving laboratory and the receiving individual are professionally qualified to accept the HMUR agent, and that the facility has adequate biosecurity to maintain the agent. This confirmation can be accomplished by using material transfer agreements or end-use agreements. An end-use agreement documents that the recipient has the facilities, personnel, and expertise to use the biological agents safely, securely, and appropriately.[1] The sending laboratory should promptly notify the receiving laboratory of the time the shipment left the sending laboratory. The sender should receive a safe-receipt notification from the final recipient. Both laboratories should be prepared to independently respond immediately if any shipment does not arrive as expected. This type of incident should also be reported to a higher authority, such as an LE agency.

5.3.6 Information Security

Sensitive information related to the security of HMURs should be marked and handled according to documented policies and procedures. Information security measures should be similar to those recommended for MRFs.

5.4 EXTREME-RISK FACILITY

Extreme-risk facilities (ERF) have extreme-risk biosecurity scenarios as their highest-risk scenarios. ERFs should implement the measures for HRFs plus the elements described below. Graded protection is critical for efficient resource allocation at ERFs because almost all of these facilities will also have lower-risk biosecurity scnarios, and high, moderate, and low risk agents do not warrant the same level of biosecurity as extreme malicious use risk agents. Because the overwhelming majority of bioscience facilities do not hold EMURs, the recommendations in this section will be limited. Any facility with EMURs should hire a professional security expert to assist in designing, implementing, and testing the laboratory biosecurity system.

5.4.1 Program Management

The biosecurity program must have support from the highest levels of management at the facility.

5.4.2 Physical Security

If a facility holds EMURs, it should designate Special Exclusion Areas where those agents are used and stored. Special Exclusion Areas should employ at least two levels of electronic access controls to ensure that only authorized personnel have access to EMURs. Special Exclusion Areas should have redundant intrusion detection systems in place. ERFs should have a centralized alarm monitoring and control system, and the access control and intrusion detection alarms from the Special Exclusion Areas should be monitored separately from the rest of the facility. All elements of this monitoring system should be kept within an Exclusion Area, data lines should have tamper indicators, and the alarm network should be isolated from all other networks. Alarms should be monitored 24 × 7 × 365 and assessed by trained security personnel.

It is highly recommended that the Special Exclusion Areas be monitored by an on-site guard force that is equipped to respond to any alarms in these areas. All alarms in Special Exclusion Areas should be routed automatically to all appropriate response forces.

Reliable communication should be established between alarm monitoring and response personnel, and a secure communications mode should be used for security incident communication. Care should be taken to preclude emergency egress pathways from channeling evacuating personnel into or through Limited, Exclusion, or Special Exclusion Areas.

5.4.3 Personnel Security

A comprehensive background investigation should be conducted to authorize personnel for work with EMURs. Access to EMURs should be limited to those who meet personnel security standards as defined by the facility that are more restrictive than what is required for access to HMURs. Furthermore, the risk associated with EMURs is such that extra precautions — for instance, a national security clearance — are generally warranted for all personnel who have authorized and unescorted access to any part of the facility where EMURs are stored or handled.

Until formal authorization is granted, the individual should not be permitted access to Special Exclusion Areas, and visitors should not be allowed in these areas without intensive escorting in place (e.g., at least two escorts for every one visitor).

5.4.4 Material Control and Accountability

A facility-wide secure inventory of EMURs should be maintained to track who works with what EMURs in which locations. Usage logs should be kept that document when the EMURs are accessed and by whom. We strongly recommend that two authorized personnel be required to access repository stocks of EMURs.

Autoclaves and other forms of decontamination equipment used for EMUR materials should reside within the same Special Exclusion Area, or equivalent, as that used for laboratory and storage purposes. If this is not possible, procedural control measures should be implemented. For decontaminating EMURs outside of Special Exclusion Areas, two individuals who are authorized to work with EMURs (or screened to an equivalent level) should be present, and independent verification of successful decontamination should be conducted. Verification procedures should be performed immediately following the EMUR decontamination process regardless of where the decontamination takes place; if it cannot be done immediately, the material should be resecured at an appropriate biosecurity level until the decontamination confirmation can be conducted.

5.4.5 Transport Security

Because transporting EMURs is a sensitive and high-risk procedure, efforts should be taken to minimize the number of times EMURs need to be transported — either within a facility or between facilities. When transport of EMURs is necessary, laboratory managers and staff should judiciously engage LE officials prior to and during the shipment. Plans should also be established to address possible safety or security incidents that might occur during the transportation of EMURs, and dry-run exercises should be conducted before the EMURs are transported.

5.4.6 Information Security

All details of the security system that is responsible for protecting EMUR materials should be carefully protected. In the U.S., security details that could provide an adversary with sufficient information to overcome or subvert the security system of facilities holding EMURs would likely be classified as national security information, stored and used in a Vault-Type Room, and only accessed by those with a need to know and the appropriate national security clearance. In other countries, this information should be held in an Exclusion Area at a minimum, preferably with additional access restrictions put in place to ensure that only those with the proper authorization have access. This information should not be taken out of the secured area unless similar protections are available during its transport and at its destination. Outdated information or copies of the information that are no longer needed should be shredded. Any copy of security documentation that needs to be removed from its secured container should be numbered and signed out by the recipient and signed back in once secured. This information should not be discussed in unsecured areas or in a manner that unauthorized personnel may overhear. Voice and data communication of this information should occur only over encrypted lines.

5.5 SUMMARY OF SPECIFIC BIOSECURITY RECOMMENDATIONS

5.5.1 Specific Biosecurity Recommendations — Low-Risk Facility

Low-risk Facilities (LRF) are defined as those facilities that have low-risk biosecurity scenarios as their highest risk. If properly implemented, laboratory biosafety measures coupled with best business practices often provide sufficient security at this level.

SECURITY COMPONENTS

Facility Type	Program Management	Physical Security	Personnel Security	Material Control and Accountability	Transport Security	Information Security
Low-Risk Facility	PI is responsible for biosecurity and ensuring that lab personnel know and understand biosafety and biosecurity measures.	One layer of restriction: property protection area (PPA) Ensure that LMURs are used and stored in PPA. Protect LMURs via good biosafety practice.	Personnel hold *low-risk* positions Conduct minimal background checks on personnel. Establish an adaptable HR system. Verify employment history and educational background. Visitors granted access by PI or authorized escort.	Standard lab records (e.g., notebooks) document the storage and use of LMUR. Accountable scientist documents stocks and use of material under his/her purview.	Document the transport of all LMURs. Specify what, when, and the recipient. PI should be aware of all LMUR transport.	Ensure that prudent policies are in place for network security, password, and email use.

5.5.2 Specific Biosecurity Recommendations — Moderate-Risk Facility

Moderate-Risk Facilities (MRF) are defined as those facilities that have moderate-risk biosecurity scenarios as their highest risk.

It is important to note the significance of a graded protection approach, as almost all MRFs will also have low- and moderate-risk assets. Therefore, an MRF should build upon the security measures of an LRF and add to them the elements described in the following table.

SECURITY COMPONENTS

Facility Type	Program Management	Physical Security	Personnel Security	Material Control and Accountability	Transport Security	Information Security
Moderate-Risk Facility	**Implement all LRF components** Establish a designated biosecurity officer. Establish policies and procedures for all components. Ensure enough funding for full biosecurity implementation. Conduct risk assessments and self-audits annually. Document all policies, procedures, and plans. Be attentive to personal issues of staff that could impact safety and security.	**Implement all LRF components** Two layers of restriction: PPA, limited area (LA) Ensure LA resides within the PPA. Store and use MMURs in LA only. Ensure access controls and intrusion detection in place in LA to ensure only authorized personnel enter and exit the area. Monitor LA regularly for signs of unauthorized access. Ensure LA has balanced strength of construction. (e.g., doors and walls are equally difficult to breach).	**Implement all LRF components** Personnel hold *moderate-risk* positions. Conduct pre-qualification check to verify identity and credentials Conduct in-depth background investigation upon hiring. Escort required until complete. Unless safety could be compromised, ensure that photo badges are worn at all time to identity personnel and authorization level. Ensure that visitors are badged and escorted. No more than 4-to-1 escort ratio in laboratory areas.	**Implement all LRF components** Identify accountable individual for all MMUR stocks. Use of lab notebooks indicate who, what, and when. Maintain seed stock inventories in secure, limited-access databases to include agent name, genus, species, strain, source, form, and responsible individual. Reflect transfers into and out of the lab and/or facility. Whenever possible, store decon equipment for MMURs in the same place as laboratory use and storage.	**Implement all LRF components** Control and document all transport of MMURs. For internal transfer, maintain internal chain-of-custody documentation. For external transfer, sending laboratory verifies capabilities of recipient laboratory. Screen all MMUR transport personnel at the same level of MMUR laboratory personnel. Maintain LA level of control until transport is underway. Establish MMUR recipient procedures.	**Implement all LRF components** Mark and handle all MMUR security information according to policies and procedures. Protect information equivalent to LA. Screen potentially sensitive information prior to public release. Establish sound policies for desktop security, passwords, and email use. Communicate sensitive information through controlled means, such as password protection and/or encryption. Train staff on operational and information security.

5.5.3 Specific Biosecurity Recommendations — High-Risk Facility

High-Risk Facilities (HRF) are defined as those facilities that have high-risk biosecurity scenarios as their highest risk.

It is important to note the significance of a graded protection approach, as almost all HRFs will also have low- and moderate-risk assets. Therefore, an HRF should build upon the security measures of an MRF and add to them the elements described in the following table.

SECURITY COMPONENTS

Facility Type	Program Management	Physical Security	Personnel Security	Material Control and Accountability	Transport Security	Information Security
High Risk Facility	**Implement all MRF components** Designated biosecurity officer ensures appropriate levels of biosecurity training. Biosecurity officer conducts self-audits and system performance tests to ensure compliance and effectiveness. Be attentive to personal issues of staff that could impact safety and security. Ensure management's authority to restrict EA from individuals who may pose safety or security risks due to behavior.	**Implement all MRF components** Three levels of restriction: PPA, LA, exclusion area (EA) Store and use HMURs in EA only. Ensure EA resides within the LA. Smaller set of individuals authorized in EA than in LA. Ensure access to EA requires unique item (e.g., electronic key) and knowledge (e.g., PIN). Implement access control and intrusion detection elements including: keys, PINs, magnetic locks, BMS, and sensors such as glass break, motion detection, or volumetric. Ensure escorted access for individuals without routine access privileges. Log-in/out upon entry/exit for both visitors and routine personnel. Security personnel periodically monitor all EA entrances and exits either in person, or via video assessment. Ensure EA doors remain locked in emergency, with RTE device for emergency egress. Employ on-site guard force or establish MOU with local guard force that specifies response time. Ensure EA has balanced strength of construction. (e.g., doors and walls are equally difficult to breach). Ensure windows are locked and alarmed or reinforced. All other entry points should be similarly secured. Place signage that indicates restricted items (cameras, PDAs, firearms) Monitor and lock EA at the end of each day.	**Implement all MRF components** Personnel hold *high-risk* positions. Limit access to HMURs to those who meet defined facility standards. (Typically higher than those established for MMURs) Conduct pre-qualification check to verify identity and credentials, as well as ensure no criminal record or terrorist affiliation. Conduct in-depth background investigation (more comprehensive than for moderate risk positions) upon hiring. Investigation should cover the preceding 10 years and address a variety of issues such as substance abuse, criminal check, credit authorization, and interviews. Reinvestigate every 5 years. Escort required until investigation is complete. Unless safety could be compromised, ensure that photo badges are worn at all time to identity personnel and authorization level. If badges are precluded by safety, other authentication methods, such as biometrics or PINs, should be used. Ensure visitors are badged and escorted. Working visitor-to-escort ratio should be no more than 1-to-1.	**Implement all MRF components** MC&A requirements for HMUR build upon MMUR. Identify accountable individual for all seed and working stocks. Biosecurity Officer should retain a record of the accountable individual responsible for each HMUR, documenting who, what, and where, as well as any transfers of HMURs. Maintain inventories of seed stocks, in secure limited-access databases to include accountable individual and/or individual doing the work, the nature of the experiment, and the status of any animals infected with HMURs, including relevant dates. Whenever possible, store decon equipment for HMURs in the same place as laboratory use or storage. When not possible, authorized HMUR personnel should remain in attendance during decontamination and verify success.	**Implement all MRF components** Implement all elements of transport control and documentation that apply to MMURs. In addition, apply additional layers of control as appropriate. Biosecurity Officer pre-approves all transport (internal and export) of HMUR. Maintain EA controls on package, or if temporary storage is necessary, package should remain in the custody of an HMUR-authorized individual. Use timely shipping methods for all external transfers. Accountable individual should verify capabilities of receiving lab and receiving individual by use of an *end-use agreement*. Promptly notify the receiving lab of time shipment left. Recipient lab should notify safe receipt. Ensure that procedures are in place for sending/receiving labs for instances when shipment does not arrive. Report such incident to higher authority.	**Implement all MRF components** Information security similar to that of Moderate Risk Facility. Address all elements mentioned under MRF. Mark and handle all HMUR security information according to policies and procedures.

5.5.4 Specific Biosecurity Recommendations — Extreme-Risk Facility

Extreme-Risk Facilities (ERF) are defined as those facilities that have extreme-risk biosecurity scenarios as their highest risk.

It is important to note the significance of a graded protection approach, as almost all ERFs will also have low-, moderate-, and high-risk assets. Therefore, an ERF should build upon the security measures of an HRF and add to them the elements described in the following table.

Facility Type	Program Management	Physical Security	Personnel Security	Material Control and Accountability	Transport Security	Information Security
Extreme Risk Facility	**Implement all HRF components** All components addressed in HRF apply. In addition, the biosecurity program must have support from the highest level of facility management.	**Implement all HRF components** Four levels of restriction: PPA, LA, EA, and Special exclusion area (SEA) Limit access control to EMUR authorized personnel only by a minimum of two electronic access controls In EA: Implement centralized alarm monitoring and control system, in which all key components are housed within the EA. Tamper indicating devices and alarm networks should be operated independently from others networks. Monitor alarms around the clock and assess alarms by trained security personnel. In SEA: Monitor by on-site response force. All alarms in SEA route automatically to all other appropriate response forces. Establish reliable communication between alarm monitoring and response force. Ensure secure line for security incident communication Emergency egress pathways should not channel personnel into LA, EA, or other SEAs.	**Implement all HRF components** Personnel hold *extreme-risk* positions. Conduct comprehensive background investigation for all personnel handling EMURs. Limit access to EMURs to those who meet defined facility standards (more restrictive than those established for HMURs). Extra precautions, such as national security clearances, are warranted for all EMUR-authorized personnel. Refuse access to SEA until all investigations are complete and all clearances are granted. Limit number of visitors in SEA, authorized by biosecurity officer. Ensure that visitors are escorted by EMUR-authorized personnel, with a ratio of two escorts to one visitor.	**Implement all HRF components** Maintain a facility-wide secure inventory of EMURs. Track individuals who work with EMURs and where. Two authorized personnel required to access EMUR stock repository. Whenever possible, store decon equipment for EMURs in the same place as laboratory use or storage. When not possible, two authorized EMUR personnel should remain in attendance during decontamination and verify success. Perform verification procedures for EMURs following the decon process, regardless of where decon takes place. If verification cannot be done immediately, EMUR should be re-secured until the decon verification can take place.	**Implement all HRF components** Minimize the amount of transport, both internal and external, for all EMURs. When necessary, lab managers and staff should engage law enforcement officials prior to and during shipment. Establish plans to address safety and/or security incidents that might occur during transport.	**Implement all HRF components** Carefully protect all security system details that protect EMURs. In the US, such information is typically considered national security information. This information is stored in a vault-type room, and accessed only by those with a 1) a need to know and 2) appropriate level of national security clearance. In other countries, security system information should be housed in the EA with additional access restrictions in place. Do not take EMUR security-related information outside of the EA without EA-level protection during transport and at final destination. Shred all outdated copies of security information. Ensure that security documentation needing to be removed is numbered and signed in/out by personnel authorized to access the information. Do not discuss security information in unsecured areas. Use only encrypted lines for voice and data communications.

REFERENCE

1. An example of a material transfer agreement can be found at: http://ott.od.nih.gov/pdfs/Bmlintrn.pdf.

6 Conclusions

The success of laboratory biosecurity starts with understanding and communicating its fundamental rationale. If the rationale is not persuasive to the bioscience community — either locally, nationally or internationally — the implementation of laboratory biosecurity will not be successful. The existence of almost every pathogen in the natural environment seemingly challenges the need to protect those same agents in laboratory settings. Yet we contend that it would be much easier for a prospective terrorist to steal a pathogen from a bioscience laboratory, where the organism has been characterized for its viability and virulence, than to locate and isolate the organism from the environment — or to create it through chemical synthesis or genetic engineering — and then conduct efficacy testing. In other words, we believe that the risk is higher that a terrorist would acquire a characterized pathogen from a laboratory than try to develop it through other means.

The difficulty of preventing a skilled scientist or technician with malevolent intent from covertly removing a dangerous pathogen from a laboratory also challenges the rationale for biosecurity. Pathogens emit so little energy that we do not have the ability with currently available technology to detect a person illicitly carrying an organism out of a laboratory. Even pervasive video surveillance, rules prohibiting one person from accessing a pathogen alone, and body searches could not definitively prevent the theft of a pathogen from a laboratory. Yet we believe that, despite these complexities, those individuals who possess and use dangerous pathogens have a responsibility to reduce the risk that those agents could be stolen to cause harm. We are also confident that reasonable and balanced laboratory biosecurity measures, which do not unnecessarily hinder the progress of science, can introduce a level of deterrence that would make it considerably more difficult — and thus less likely — for a person to steal a dangerous pathogen from a legitimate bioscience laboratory.

We acknowledge that the risk of naturally occurring pandemic infectious disease is higher today than catastrophic bioterrorism, but we are concerned that the bioterrorism risk will increase over time. The rapid advances in biotechnology, which show great promise for improving human and agricultural health in the future, will also gradually lower the technical barriers to bioterrorism. As biotechnology advances, it will become easier and easier, and cheaper and cheaper, for more people to misuse biology to cause harm. In other words, without laboratory biosecurity systems in place, bioscience laboratories will become more vulnerable to malicious exploitation over time. Laboratory biosecurity is as much about reducing the risk that legitimate bioscience facilities could be sources of pathogens and toxins for malicious use in the future as today.

There are also other specific benefits of laboratory biosecurity that this book has emphasized. Laboratory biosecurity can protect valuable research and commercial assets. It is for this reason that, in general, the private biotechnology industry is more comfortable with the concepts of laboratory biosecurity than the government and academic laboratories. Laboratory biosecurity also provides assurance to citizens and investors about the protection of dangerous pathogens and toxins. Citizens, particularly those who live and work in the vicinity of a biological containment laboratory, sometimes worry about the management and oversight of infectious disease facilities, fearing that disease-causing agents will be accidentally or deliberately released into their community. Investors in biotechnology ventures do not want to incur excessive liabilities because their facilities may be perceived as unsafe or not secure. By establishing consistent and transparent laboratory biosecurity practices, both citizens and investors can be assured that appropriate steps have been taken to reduce the risk that dangerous pathogens and toxins in these facilities will be mishandled or misused.

Yet another benefit of laboratory biosecurity is that it complements the laboratory biosafety agenda of preventing disease in people, animals, and plants and minimizing the risk of worker injury. A comprehensive biosafety program provides the foundation for effective laboratory biosecurity. A laboratory that lacks a functional biosafety process will not be able to implement a thorough laboratory biosecurity system. At the same time, laboratory biosecurity reinforces and strengthens the biosafety culture, helping to ensure the containment of hazardous infectious substances in laboratories. Fundamentally, laboratory biosafety and biosecurity both focus on managing the risks of working with infectious disease agents in bioscience laboratories.

This book has also articulated that the similarities of laboratory biosafety and biosecurity extend beyond objectives to include methodology and implementation. For both biosafety and biosecurity, the first step in managing risk is defining the problem through a systematic risk assessment. The biosafety and biosecurity risk assessments may be similar but they should be distinctly different processes. Risk managers must then decide which risks are acceptable and which risks are unacceptable, recognizing that risk cannot be reduced to zero. This risk decision defines the performance objectives and specific strategies of the controls and system that need to be implemented to manage those risks.

As with biosafety, a combination of engineering and procedural controls can achieve the objectives of laboratory biosecurity. In many cases, procedural controls can be as effective as engineered controls. Procedural controls rely on individuals behaving in a specified manner, whereas engineering controls are designed to prevent individuals from behaving in undesired ways. For example, a facility could implement procedures that require laboratory workers to sign a log book upon entry and exit of a containment suite, or the facility could use proximity card readers that would detect when an individual entered or exited the laboratory. Both of these systems meet the objective of documenting which individuals are in the laboratory at a given time. The log book is inexpensive and easy to implement, but relies on the compliance of staff. The proximity card system is more expensive and requires a more cumbersome installation process, yet it reduces the staff's burden of signing a log book, and it is more difficult to defeat. However, even the more sophisticated system will not be effective if it malfunctions or if a staff member holds the door open for someone who does not have the proper permission or proximity card.

Thus, even with extensive engineering controls, maintaining safety and security in biological containment facilities depends primarily on the individuals who work there. The laboratory-acquired infections of SARS in 2003 occurred in BSL3 and BSL4 laboratories. A WHO investigation attributed those infections to negligent program management (e.g., poor laboratory practices, insufficient training). Similarly, even the most sophisticated security systems can be circumvented if the people with access to dangerous pathogens are not trustworthy, reliable, or trained to abide by the security protocols.

Successful biosecurity ultimately depends on the ability of the institution and bioscience community at large to create a culture of responsibility that reflects an awareness of the potential misuse of biological materials and an acceptance of biosecurity practices.[1] Such a transition, similar to the acceptance of laboratory biosafety several decades ago, will be important as laboratory biosecurity spreads across the globe. In the past, it was not uncommon for scientists to use mouth pipettes, a practice that today is recognized as so unsafe that it is almost nonexistent in laboratories. Although laboratory biosafety as an area of expertise has existed for more than 50 years, the first definitive texts in the field — the CDC and NIH's *Biosafety in Microbiological and Biomedical Laboratories* and the WHO's *Laboratory Biosafety Manual* — did not appear in print until the early 1980s. Each of those texts has been revised multiple times since its first publication.[2] The professional field of laboratory biosafety is still evolving today, partly as a result of technological innovation but also because of the gradual acceptance by laboratory managers, scientists, and technicians of the importance of working safely with biological agents.

The field of laboratory biosecurity finds itself in a similar circumstance today as laboratory biosafety several decades ago. As indicated in the Introduction, the material presented in this book will inevitably be subject to revision and improvement in the future. However, part of the reason that biosecurity methods remain somewhat conceptual is that biosecurity is still not widely practiced. Bioscience laboratories, even those that work with dangerous pathogens and toxins, regularly leave doors unlocked, do not conduct background checks on their personnel, and allow visitors open access to their facilities. In the not-too-distant future, perhaps these practices also will be rare as laboratory staff and management perceive biosecurity as a fundamental responsibility for those individuals and institutions that work with pathogens or toxins.

Developing this biosecurity culture starts with outreach to those who work at bioscience institutions to ensure that they understand the rationale for laboratory biosecurity. Several systematic surveys of laboratory-acquired infections that documented laboratory-based safety risks helped establish the intellectual foundation for implementing laboratory biosafety.[3] We do not currently have a similar comprehensive public accounting of laboratory biosecurity incidents. Nevertheless, policy makers and laboratory managers, even in the absence of this data, need to make an effort to educate the bioscience community on the risks. Basic awareness training then needs to be sustained through institutional commitment. Allocating adequate resources and providing regular training for staff are two mechanisms that can help build a biosecurity culture by demonstrating the importance of laboratory biosecurity to institutional management.

We are optimistic that laboratory biosecurity will evolve more quickly than laboratory biosafety. Not only are there now many countries around the world with legislation or regulations that require some form of laboratory biosecurity but also the concept of laboratory biosecurity can benefit from all the progress that biosafety has made over the years. Today there is a rapidly expanding network of professional associations dedicated to advancing laboratory biosafety. The American Biological Safety Association was established in 1984,[4] the European Biosafety Association was founded in 1996,[5] the Brazilian Biosafety Association was created in 1999,[6] and the Asia-Pacific Biosafety Association was founded in 2005. Biosafety associations also exist in Taiwan and Japan. Although these associations are dedicated to advancing the practice of laboratory biosafety, many of them have embraced and actively promote the concepts of laboratory biosecurity.

As the Introduction emphasized, laboratory biosecurity is an international enterprise. Laboratories that work with dangerous pathogens exist throughout the developed, developing, and underdeveloped world. Protecting dangerous pathogens from theft in one or a few countries, and not internationally, will only have a marginal effect on the bioterrorist threat. The same is true for biosafety. Fortunately, there is growing international attention focused on the topic. For example, the International Committee of the Red Cross (ICRC) has called on governments to strengthen their commitment to international legal norms that prohibit the hostile uses of biological agents and safeguard potentially dangerous agents, technology, and knowledge.[7] In 2004, the Organisation for Economic Development and Cooperation (OECD) initiated a biosecurity program to promote responsible stewardship in the biosciences and avoid the potential abuse of research and resources.[8] The following year, the International Criminal Police Organization (Interpol) launched an initiative to enhance the controls on, and prevent the malicious use of, pathogens and toxins.[9]

These international organizations demonstrate that the field of laboratory biosecurity is fundamentally multidisciplinary. Although it directly affects the work of laboratory bioscientists, laboratory biosecurity has become vital for the policymaking, national-security, law-enforcement, and emergency-response communities. Laboratory biosecurity has also revealed that it is increasingly essential for these various communities to communicate with each other in a way they never have before. There is a need for scientists to understand security issues and assist the national and international security communities develop policies, and there is a need for the national security communities to become much more familiar with the practice of bioscience. Laboratory biosecurity provides an excellent interface for this multidisciplinary communication. We hope this book — which has attempted to span the disciplines of science and security — will be useful to all of these communities as they deliberate on the implications as well as the implementation of laboratory biosecurity.

REFERENCES

1. Moodie, M., 2003, Chemical and Biological Arms Control Institute, Special Report 5, Reducing the Biological Threat: New Thinking, New Approaches, Washington, D.C.
2. U.S. Department of Health and Human Services, Centers for Disease Control and Prevention, and National Institutes of Health, 1999, *Biosafety in Microbiological and Biomedical Laboratories (BMBL)*, 4th ed., Washington, D.C.; World Health Organization, 2004, *Laboratory Biosafety Manual*, 3rd ed., Geneva: WHO.
3. Barbeito, M.S. and Kruse, R.H., A History of the American Biological Safety Association, http://www.absa.org/abohist1.html.
4. http://www.absa.org.
5. http://www.ebsaweb.eu.
6. http://www.anbio.org.br.
7. International Committee of the Red Cross, 2002, Biotechnology, Weapons and Humanity: Appeal of the International Committee of the Red Cross, Geneva: ICRC.
8. http://www.biosecuritycodes.org/.

List of Acronyms

ABSA American Biological Safety Association
ABSL Animal Biosafety Level
AHP Analytic Hierarchy Process
APHIS Animal and Plant Health Inspection Service
BMBL Biosafety in Microbiological and Biomedical Laboratories (CDC/NIH)
BMS Balanced Magnetic Switch
BSC Biosafety Cabinet
BSL Biosafety Level
BSO Biosafety Officer
BW Biological Weapon
BWC Bacteriological (Biological) and Toxin Weapons Convention
CAS Central Alarm Station
CCTV Closed Circuit Television
CDC Centers for Disease Control and Prevention
CFR Code of Federal Regulation
DC District of Columbia
DGAC Dangerous Goods Advisory Council
DOC Department of Commerce
DOJ Department of Justice
EA Exclusion Area
EAP Employee Assistance Program
EMUR Extreme Malicious Use Risk
EPA Environmental Protection Agency
ERF Extreme-Risk Facility
FBI Federal Bureau of Investigation
FMD Foot and Mouth Disease
FOIA Freedom of Information Act
GAO Government Accountability Office (U.S.)
GMO Genetically Modified Organism
HEPA High Efficiency Particulate Air (filter)
HMUR High Malicious Use Risk
HR Human Resources
HRF High-Risk Facility
HSPD Homeland Security Presidential Directive
IATA International Air Transport Association
IBC Institutional Biosafety Committee
IBTR International Biological Threat Reduction Program
ICAO International Civil Aviation Organization
ICRC International Committee of the Red Cross
Interpol International Criminal Police Organization

LA Limited Area
LAI Laboratory-Associated Infection
LAN Local Area Network
LBM *Laboratory Biosafety Manual* (WHO)
LE Law Enforcement
LMUR Low Malicious Use Risk
LRF Low-Risk Facility
MC&A Material Control and Accountability
MMUR Moderate Malicious Use Risk
MOU Memorandum of Understanding
MRF Moderate-Risk Facility
NIH National Institutes of Health
NIMS National Incident Management System
NRP National Response Plan
NSABB National Science Advisory Board for Biosecurity
OECD Organisation for Economic Co-operation and Development
OIE Office Internationale Epizooties (World Organization for Animal Health)
PDA Personal Digital Assistant
PDD Presidential Decision Directives
PI Principal Investigator
PIN Personal Identification Number
PPA Property Protection Area
PPE Personal Protective Equipment
RO Responsible Official
RTE Request-to-Exit
SAP Select Agent Program
SARS Severe Acute Respiratory Syndrome
SBU Sensitive but Unclassified
SEA Special Exclusion Area
SNL Sandia National Laboratories
SOP Standard Operating Procedure
TID Tamper Indicating Device
U.N. United Nations
UNSCR United Nations Security Council Resolution
U.S. United States
vLAN Virtual LAN
VPN Virtual Private Network
WHO World Health Organization
WMD Weapons of Mass Destruction

Glossary

Adversary A person performing malevolent acts; an adversary may be an insider or an outsider.

Asset Item of value from the perspective of either the institution or a potential adversary. The item may be dangerous, rare, valuable, or otherwise hard to replace, or may cause an unacceptable delay in mission execution if lost.

Extreme Malicious Use Risk Group Biological agents that have a similar task complexity and potential consequences as high malicious use risk group agents (low task complexity to be used maliciously and moderate- to high-potential consequences if successfully used maliciously), except they receive a higher classification because they are not found in nature.

High Malicious Use Risk Group Biological agents that have low task complexity to be used maliciously and moderate- to high-potential consequences if successfully used maliciously. These biological agents would not be difficult to deploy as a weapon, resulting in national or international consequences with moderate-to-high casualties and/or economic damage and the potential to cause mass panic and significant social disruption.

Insider A person with authorized access. This person, by reason of official duties, has knowledge of operations and/or access that would significantly enhance the likelihood of successful bypass or defeat of biosecurity measures, should that person attempt such an action.

Laboratory Biosafety A set of preventive measures designed to reduce the risk of accidental exposure or release.

Laboratory Biosecurity A set of preventive measures designed to reduce the risk of intentional removal (theft).

Low Malicious Use Risk Group Biological agents that have high task complexity to be used maliciously and low potential consequences if successfully used maliciously. They may be difficult to acquire, difficult to produce in sufficient quantities, and unsuitable for dissemination. They would be expected to have low population impact, inflict little economic damage, and have a low psychological impact on the general population.

Malicious Use Risk Group Classification scheme for describing the relative risks that biological agents can be used for bioterrorism or other malicious acts. Assigning a biological agent to a malicious use risk group is the starting point for a biosecurity risk assessment.

Moderate Malicious Use Risk Group Biological agents that have moderate task complexity to be used maliciously and moderate potential consequences if successfully used maliciously. Many of the steps associated with successfully deploying these agents will be relatively easy, but perhaps one or two of the critical steps are difficult. They would be expected to cause localized consequences with low-to-moderate casualties, moderate-to-significant economic damage, and potentially cause pervasive anxiety.

Notional Adversary A hypothetical adversary with a well-defined set of assumptions used for the risk assessment. Insider, single terrorist outsider, and outsider terrorist group are all examples of notional adversaries.

Outsider A person that does not have authorized access. This person does not have official business with the facility.

Potential Consequences Potential impact of an undesired event on society and the institution.

Restricted Area A specifically defined area to which access is controlled.

Risk The likelihood an adverse event will occur.

Risk Assessment A systematic, structured process for analysis and determination of risk.

Safety Risk Group Classification scheme for describing the relative safety risks that biological agents can pose to the individual and the community. Assigning a biological agent to a safety risk group is the starting point for a biosafety risk assessment.

Scenario Undesired event; a combination of an asset, an adversary, and an action.

Task Complexity The ease or difficulty in executing the undesired event. For biological assets, task complexity may be viewed as the difficulty associated with turning a biological agent into a weapon.

Threat Potential The potential that an adversary may pose a threat to an asset. Threat potential is used as a proxy for probability.

Unauthorized Person Person not authorized to have access to specific information, materials, or areas.

Appendix A

Vulnerability Assessment Questionnaires

A.1 SITE EVALUATION QUESTIONNAIRE

Institute:		
Campus:		
Subject: **Site/Perimeter Conditions**		
Completed by:		Date:
No.	**Security Issue**	**Answers/Comments**
1	Does a fence or wall surround the perimeter? If so, what type, height, and to what degree does it surround the perimeter, and what is its condition?	
2	Provide a sketch of the site features, including the perimeter fencing, access roads, buildings, and high-priority assets (include RR track, power station, access gates, etc.).	
3	Is access to the site controlled? If so, describe.	
4	Do visitors routinely sign in to all areas where required? Are escorts required? If so, where?	
5	Are all visitors required to wear badges?	
6	Is exterior lighting adequate?	
7	Does security equipment have emergency power?	
8	Describe the condition of the site (is the site overgrown, etc.).	
9	Describe any deterrents associated with the perimeter.	
10	Are there any topographic restrictions to access to the site? If so, describe.	

11	Are security codes, card keys, and mechanical keys closely controlled and changed when appropriate?	
12	Does intrusion detection exist? If so, describe.	
13	Is closed circuit television (CCTV) available to the facility? If so, describe the type, location, condition, and lighting.	
14	How often are the physical and electronic security systems maintained and tested?	
15	Describe the condition of the buildings.	
16	Do ground-level windows exist? If so, are they locked? Or equipped with security bars/features?	
17	Are there other openings/portals at ground level? If yes, describe.	
19	What is the economic condition of the surrounding area?	
20	Is there any organized political opposition to the institute's activities?	

Subject: **Site Threat Assessment Questionnaire**		
Completed by:		Date:
No.	**Threat Issue**	**Answers/Comments (Please include timelines when available)**
1	Has there ever been any indication that terrorists are targeting the site?	
2	Has the site had any problem with animal rights or other activist groups?	
3	Have there been any other forms of picketing/protest?	
4	Are labor relations healthy? Is the staff well paid? Well respected?	
5	Are there any local "characters" causing incidents?	
6	Is there any racial and/or religious tension within the staff and/or larger community?	
7	Has the institute experienced any petty crime or vandalism? Describe.	

8	Has the institute experienced any incidents associated with disgruntled employees?	
9	Have there been instances of violence in the workplace?	
10	Who has arrest authority at the institute?	
11	What is the frequency of law enforcement patrols of the site?	
12	Does the institute perform any kind of background investigation on institute staff?	

Subject: **Infrastructure**		
Completed by:		Date:
No.	**Security Issue**	**Answers/Comments**
1	Who owns and maintains each on-site utility (water, gas, phone, electrical, etc.)?	
2	Describe the reliability of the utilities.	
3	Is there redundant capability for each utility?	
4	Describe the construction (above ground, below ground, etc.).	
5	How are the utilities accessed?	

Subject: **General Administration**		
Completed by:		Date:
No.	**Security Issue**	**Answers/Comments**
1	Are employee background investigations conducted?	
2	Are reinvestigations conducted?	
3	Are confidentiality agreements required?	
4	Are medical examinations conducted?	
5	What type of training is provided prior to employment (fire, safety, security, biology)?	
6	Are scientists ever required to use a two-person rule?	
7	Is there a key-control program in place?	
8	Are badges required?	
9	Are guests and visitors escorted?	

10	Are follow-up investigations conducted on incident reports?	
11	Are deliveries received at a central location, and are there manifests for all deliveries?	
12	Are HAZMAT policies and procedures in place?	
13	Do you exercise safety/security incident response?	
14	Are information sheets available to assist in the event of a threat (phone, bomb, etc.)?	
15	What governing agencies have influence over the type of operations that occur on-site?	
16	Who funds the activities at this facility?	
17	Are there any new buildings planned?	
18	IT security protocols? Passwords? Screen savers? Firewalls, etc.? Encryption capability?	

A.2 LAW ENFORCEMENT QUESTIONNAIRE

Institute:		
Campus:		
Subject: **Local Law Enforcement**		
Completed by:		Date:
No.	**Threat Issue**	**Answers/Comments**
1	Has there been any terrorist (domestic or international) activity in the region, or indication that terrorists are targeting the site?	
2	What types of crime exist near the institution?	
3	Acts of political/religious extremism?	
4	Animal rights or other activists?	
5	Other forms of picketing/protest? Labor relations?	
6	Any local "characters" causing incidents?	
7	Racial/religious tension?	
10	Petty crime? Vandalism?	
11	Is a historical record of the criminal activity in the area available? (Within the last 6 months? Within 10 miles of the institute?)	

12	Incidents with disgruntled employees? Or other instances of being called out to the institute?	
13	Does LE have arrest authority at the institute?	
14	Are mock exercises with the institute conducted? If so, describe.	
15	Does a Memorandum of Understanding/ Agreement exist with the institute regarding emergency response?	
16	Has a chain of command been established between local LE response and the institute?	
17	Is there any direct alarm communication between the institute and the local LE? If not, how would LE be summoned? Is the communication system reliable?	
18	How often is the scientific institute briefed on threats (daily, weekly, monthly, for cause)?	
19	How many LE personnel are available to respond to the institute on each shift?	
20	What is the average response time to the institute? Any logistical issues with the route?	
21	What is the frequency of LE patrol of the site?	
22	What communication systems are available? What is the backup communication system?	
23	Are communication systems compatible with the institute security force?	
24	What are the roles and responsibilities of each organization that would be involved in identifying a threat to the institute (biological in general) and responding to it?	
25	Is a map available of the areas patrolled?	
26	Are there any jurisdictional issues with other LE agencies?	
27	Does LE help the institute in response force planning, transport of dangerous materials, or any other form of security service?	
28	Do LE personnel know what kinds of dangerous pathogens are held at the institute, where they are held, and who is responsible for them?	
29	Are LE personnel familiar with biological agents?	

30	Does LE have technical assistance available to answer questions regarding biological or chemical materials?	
31	Would LE know who to talk to at the institute in the event of a specific threat?	
32	Would LE know how to enter a bioscience facility safely?	
33	What kinds of training do the LE response personnel receive?	
34	If there is an on-site institute guard force, is there any communication or training between on-site guards and LE response personnel?	
35	Are there legal restrictions to how LE responds to an alarm/call for armed adversaries at the institute?	
36	Is there an established "Use of Force" policy? If so, what are the parameters?	

A.3 GUARD FORCE QUESTIONNAIRE

Institute:		
Campus:		
Subject: **Guard Force**		
Completed by:		Date:
No.	**Security Issue**	**Answers/Comments**
1	Describe the guard force's mission objectives.	
2	Provide an organizational chart.	
3	Who has the highest level of decision-making authority within the guard force?	
4	Who does the guard force report to?	
5	Who within the guard force organization/chain of command reports to the head of the science institute?	
6	Describe the guard force's duties for routine and emergency situations.	
7	How do these activities differ depending on shift?	

8	What level of background investigation is required for guard force employment?	Police: Contract guard force:
10	How often are employment reinvestigations conducted?	
11	Are confidentiality agreements required?	
12	Are medical examinations conducted?	
13	What type of emergency response training is provided? Facility-specific training?	
14	Under what conditions is guard force assistance requested during employee terminations?	
15	Is the emergency response plan current? Other security plans?	
16	Are mock exercises conducted? If so, describe.	
17	Does a Memorandum of Understanding exist with the local LE?	
18	Has a chain of command been established between the institutional guard force and local LE?	
19	How often is the guard force briefed on threats (daily, weekly, monthly)?	
21	How many guard force personnel are on duty for each shift?	
22	How many guard force personnel respond to normal security-related events? What if something more serious is going on?	
23	During a response to a high-level intrusion, are other security events being monitored and/or responded to?	
24	What are the procedures for responding to restricted areas?	
25	Are there any special techniques used when a guard encounters a potentially contaminated suspect?	
26	Is there an established "Use of Force" policy? If so, what are the parameters?	
28	Are response policies in place for various scenarios (i.e., protesters vs. unauthorized access)?	
29	What is the response time for outside agency assistance?	

30	What is the average on-site response time?	
31	Do response procedures differentiate between an inside threat and an outside adversary? If so, in what way?	
32	How are calls or dispatches prioritized?	
33	How are incidents reported?	
34	Who receives and reviews incident reports?	
36	What types of weapons are carried (batons, knives, guns, etc.)?	
37	What types of weapons are available? How are they controlled?	
38	List the means of transportation (vehicle, motorcycle, foot, etc.).	
39	What is the frequency of patrol to various parts of the facility?	
40	What areas are prohibited to guards without an escort?	
41	What are the methods of communication?	
42	Are any on-site communication modes encrypted?	
43	What are backup methods of communication?	
44	Are the guards immunized from the pathogens in the laboratories?	
45	Is a map available of the areas patrolled? Does it indicate areas with dangerous pathogens or other hazards?	
46	Are training exercises conducted? If so, how often and what type?	
47	Is there a direct alarm between the institute and the guard force? How is local LE summoned?	

Appendix B

Example Biosecurity Risk Assessment Methodology

Biosecurity risk assessment is a tool to help the risk manager implement appropriate biosecurity risk reduction measures. There are many ways to achieve this goal. In this section, we discuss one methodology that we have successfully used at bioscience facilities both in the U.S. and internationally. We sought to develop a standard, consistent set of criteria that could be used to conduct a systematic assessment of the relative biosecurity risks.

This methodology is based on a mechanism for analyzing risk known as *Multiple Criteria Data Analysis*. It uses a hierarchical value tree of criteria, which "links technological performance information with decision criteria and weightings elicited from decision-makers, allowing visualization and quantification of the trade-offs involved in the decision-making process."[1] The methodology relies on a qualitative evaluation of the criteria, which results in a "score." To acknowledge that not all criteria will contribute equally to risk, each criterion can have an associated weight. There are many mechanisms for establishing the relative importance (weight) of each criterion. We use a widely accepted approach for expert elicitation known as the Analytic Hierarchy Process (AHP) as implemented by the Expert Choice™ decision support software tool. This algorithm[2] is used extensively in the industry for making complex multivariate decisions with inputs from multiple stakeholders. AHP uses a quantitative comparison method that is based on pairwise comparisons of decision criteria, rather than utility and weighting functions. The key AHP assumption is that humans are much better at making relative judgments than absolute judgments.

A weighted average is calculated to the values of the Threat Potential and the Consequences variables. The overall biosecurity risk is a function of these two variables (Equation B.1).

Equation B.1 Biosecurity Risk

$$\text{Biosecurity Risk} = (T)*(C)$$

where

T = Threat potential
C = Consequences

The relationship between these measures results in the relative risks for the series of scenarios that are relevant to a specific facility with its particular biological assets. Appropriate personnel can then decide which scenarios represent acceptable risks and which represent unacceptable risks.

B.1 ASSET ASSESSMENT (STEP 1A)

As discussed in Chapter 2, we begin by identifying and evaluating the pathogens and toxins at the facility (asset assessment). An agent's baseline risk is a function of the complexity of the task of using the agent maliciously and the maximum credible potential consequences of malicious use.

Equation B.2 Agent risk

$$\text{Agent risk} = (\text{TC}) * (\text{C})$$

where

TC = Agent task complexity
C = Consequences

First, a facility's assets must be identified. This can be accomplished through a variety of mechanisms, such as holding a meeting of stakeholders, interviewing the institution's staff, and touring the facility. Assets include anything of value to the institution or to an adversary; for simplicity we limit the discussion here to biological agents. Although it is important to be inclusive during the identification step so as to ensure nothing is overlooked, even a small facility will likely have a large number of assets that will be identified. After completing the asset assessment, it will probably be determined that many of these assets pose an insignificant risk and do not need to be carried through the assessment.

An evaluation of biological agents begins with an examination of their biochemical properties. Characterizing the task complexity associated with using a biological agent maliciously can be grouped into three criteria with associated metrics (Equation B.3). We have structured the metrics so that higher scores are indicative of a higher risk.

Equation B.3 Agent Task Complexity

$$\text{TC} = \text{TC}_{\text{A}}\text{w}_{\text{A}} + \text{TC}_{\text{De}}\text{w}_{\text{De}} + \text{TC}_{\text{Di}}\text{w}_{\text{Di}}$$

where

TC_{A} = Acquisition task complexity
TC_{DE} = Development task complexity
TC_{Di} = Dissemination task complexity
w_{A} = Weight of TC_{A} criterion
w_{De} = Weight of TC_{De} criterion
w_{Di} = Weight of TC_{Di} criterion

1. TC_A: Includes the difficulty of acquiring the agent, with consideration given to how many other pathways are possible sources for the agent in addition to the facility being assessed. Sources of agent include isolation of a virulent strain from the natural environment, theft from a legitimate facility, or creation of dangerous material through *de novo* synthesis or molecular modification.

0	Easily acquired by any or all avenues; not regulated or synthetic, and natural isolation pathway is trivial (e.g., within skill set of an undergraduate biology student), or suitable natural source globally distributed
1	Limited regulation (rarely restricted) globally, or suitable natural source available nationally, or synthetic pathway and natural isolation is easy (e.g., within skill set of experienced technician, master's student)
2	Highly regulated agent globally, or limited suitable natural sources or synthetic pathway or natural isolation pathway is difficult
3	Current facility 1 of ~25 facilities with the agent worldwide, and suitable natural sources are scarce and synthetic, or natural isolation pathway requires advanced skills (never been achieved for agent of interest but has been for a related agent)
4	Current facility sole, or almost sole, source; agent has been eradicated from nature; synthetic pathway would be ground-breaking

2. TC_{De}: Includes the difficulty of processing the agent into a suitable quantity in a suitable form for most appropriate dissemination pathway (i.e., pathway that leads to maximum credible consequences). This includes consideration of the level of testing and evaluation required, the difficulty of covert production, and storage issues.

0	Likely to require extensive research and development prior to consideration as a biological weapons agent; covert production requires sophisticated capabilities; highly unstable preparation (storage of only a few days, and will not survive dissemination stresses without advanced processing)
1	Requires extensive testing and evaluation prior to use; covert production requires extremely sophisticated capabilities (e.g., BSL4); unstable preparations (storage is difficult — unknown processes and limited viability postdissemination — l < 1 h)
2	Requires moderate investment in testing and evaluation; covert production requires advanced capabilities (e.g., BSL3, prophylaxis); relatively unstable preparations (dry preparations very difficult; liquid preparations may be stored for months)
3	Requires minimal investment in testing and evaluation; covert production requires general technical capabilities and/or protocols (e.g., BSL2); preparations are stable (can be easily stored for weeks and is viable postdissemination for hours to days).
4	Requires almost no testing or evaluation to work reliably; covert production requires minimal or no technical capabilities and protocols; preparations are extremely stable (long-term storage is straightforward; and the material is viable postdissemination for weeks or longer)

3. TC_{Di}: Includes the difficulty of disseminating the agent to cause harm. This depends on the amount of agent required for the chosen route of dissemination (e.g., dermal, inhalation, oral, or vector-borne); the analysis assumes that the most appropriate dissemination pathway is selected (i.e., the one that leads to maximum credible consequences). The stability of the agent in the environment after dissemination also impacts the difficulty in achieving successful dissemination.

0	Successful dissemination would be very difficult because the agent is known or presumed to not be infectious or toxic and/or the agent is highly unstable in the environment.
1	Successful dissemination would be difficult because the agent has or is assumed to have a low infectivity or toxicity (ID50 or LD50 > 50,000; Toxin LD_{50} > 5000 μg/70 kg) and/or it is unstable in the environment.
2	Successful dissemination would be moderately difficult to achieve because the agent has or is assumed to have a moderate infectivity or toxicity and/or it is only somewhat stable in the environment.
3	Successful dissemination would be easy because the agent has or is assumed to have a high infectivity or toxicity (ID50 or LD50 100–1000; Toxin LD50 0.1–100 μg/70 kg) and/or it is stable in the environment.
4	Successful dissemination would be very easy because the agent has or is assumed to have an extremely high infectivity or toxicity (ID50 or LD50 < 100; Toxin LD50 < 0.1 μg/70 kg) and/or the agent is extremely stable in the environment.

Characterizing the task complexity associated with using the biological agent maliciously is only one component of the agent assessment. The potential consequences must also be assessed with a standardized set of metrics. Here we examine three types of consequences. This assessment assumes that the adversary could achieve the maximum credible event, which is a valid assumption for comparing relative risks.

Equation B.4 Consequences

$$C = C_p w_p + C_e w_e + C_{psy} w_{psy} + C_{op} w_{op}$$

where

C_p = Population impact
C_E = Economic impact
C_{psy} = Psychological impact
C_{op} = Operational impact
w_p = Weight of C_p criterion
w_e = Weight of C_e criterion
w_{psy} = Weight of C_{psy} criterion
w_{op} = Weight of C_{op} criterion

1. C_p: Population impact can be further subdivided into criteria for transmissibility, morbidity, mortality, preexposure countermeasures, and postexposure countermeasures (Equation B.5).

Equation B.5 Population Impact

$$C_p = C_t w_t + C_{morb} w_{morb} + C_{mort} w_{mort} + C_{pre} w_{pre} + C_{post} w_{post}$$

where

C_t = Transmissibility
C_{morb} = Morbidity
C_{mort} = Mortaltiy
C_{pre} = Preexposure countermeasures
C_{pre} = Postexposure countermeasures
w_t = Weight of C_t criterion
w_{morb} = Weight of C_{morb} criterion
w_{mort} = Weight of C_{mort} criterion
w_{pre} = Weight of C_{pre} criterion
w_{post} = Weight of C_{post} criterion

a. C_t: An agent's transmissibility provides insight into how contagious the disease may be.

0	No potential for transmission
1	Transmission primarily via parenteral or sexual exposure (direct contact)
2	Transmission primarily via blood or contaminated objects (close contact)
3	Transmission via large droplets (e.g., casual contact of < 1 m)
4	Transmission via aerosols (e.g., remote contact of > 1 m)

b. C_{morb}: The morbidity rate reflects the percentage of people that contract the disease upon exposure to the agent, assuming an unvaccinated population.

0	The agent has a low rate (0–50%) of morbidity but would not be likely to require any treatment.
1	The agent has a low rate (0–50%) of morbidity that would most likely require outpatient treatment.
2	The agent has a low rate (0–50%) of morbidity that would most likely require hospitalization.
3	The agent has a high rate of morbidity (50–100%) that would most likely require outpatient treatment.
4	The agent has a high rate of morbidity (50–100%) that would most likely require hospitalization.

c. C_{mort}: The mortality rate reflects the percentage of people to die following a symptomatic infectious dose or toxicity, assuming they are unvaccinated and untreated.

0	Resulting disease does not typically result in fatalities in healthy adults (expected mortality rate of <1%)
1	Resulting disease has a low fatality rate (expected mortality rate of 1–15%)
2	Resulting disease has a moderate fatality rate (expected mortality rate of 15–50%)
3	Resulting disease has a high fatality rate (expected mortality rate of 50–90%)
4	Resulting disease is almost uniformly fatal (expected mortality rate of > 90%)

d. C_{pre}: Criteria for preexposure countermeasures reflect the availability and effectiveness of such countermeasures (e.g., vaccines).

0	Preexposure countermeasures for the disease caused by the agent are highly effective and readily available (95% or more effective and no significant drawbacks exist).
1	Preexposure countermeasures for the disease caused by the agent are moderately effective (over 90% effective but hurdles to administration exist, including side effects and/or somewhat burdensome inoculation procedures).
2	Preexposure countermeasures for the disease caused by the agent are minimally effective (less than 90% effective and/or significant hurdles to administration exist including IND drugs, dangerous side effects, or burdensome inoculation procedures).
3	Preexposure countermeasures for the disease caused by the agent do not exist but are under development.
4	No preexposure countermeasures exist for the disease caused by the agent.

e. C_{post}: Criteria for postexposure countermeasures reflect the availability and effectiveness of such interventions (e.g., antibiotic drugs).

0	Postexposure countermeasures for the disease caused by the agent are highly effective (very effective in all but the latest stages in the course of the disease).
1	Postexposure countermeasures for the disease caused by the agent are moderately effective (effective, but best results seen when applied early in the course of the disease).
2	Postexposure countermeasures for the disease caused by the agent are minimally effective (rarely effective, even when applied extremely early in the course of the disease).
3	Postexposure countermeasures for the disease caused by the agent are under development.
4	No postexposure countermeasures exist for the disease caused by the agent.

2. C_e: Economic impact seeks to characterize the impact to the national economy directly associated with the malicious use of the biological agent.

0	No economic impact.
1	No fluctuation in financial markets or international trade. Economy able to rebound and recoup losses within a matter of days. Minor financial impact associated with recovery.
2	Small fluctuation in financial markets and/or international trade. Economy able to rebound and recoup losses within weeks. Minor ongoing increase in the cost of doing business.
3	Moderate fluctuation in financial markets and/or international trade. Economy may require government intervention to recover. Moderate increase in the cost of doing business due to increased security measures imposed by either government or industry.
4	Major global and national impact on financial markets and international trade. Immediate government intervention required to stabilize the economy. Substantial cost associated with recovery. Significant increase in the cost of doing business causing various segments of the economy to go into bankruptcy or to cease operations.

3. C_{psy}: Psychological impact estimates how the public's behavior or the public's perception of risk is affected.

0	An attack would have no significant impact on the behavior of the public.
1	An attack would have a low impact: scattered anxiety among the public, no social disruption.
2	An attack would have a moderate impact: pervasive anxiety among the public, with minimal social disruption (small-scale closures of public infrastructure, such as schools, local transportation systems, and government offices).
3	An attack would have a high impact: mass panic among the pubic, with significant social disruption but the continuation of essential societal functions (widespread closures of public infrastructure, such as schools, regional transportation networks, some government services halted), moderate threat to national security.
4	An attack would have an extreme impact: mass panic among the public, with extreme social disruption (elements of public infrastructure cease to function), significant threat to national security.

4. C_{op}: Operational impact is not included in scenarios for theft and malicious-use of biological agents. It is unlikely that the facility will be attributed as the source of the biological agent used in a bioterrorism incident and, as a result, the facility-specific impacts will be minimal.

Evaluating agents against these task complexity and consequence criteria allows the risk assessor to assign the agents to malicious use risk groups. This determination will serve as the starting point for the facility-specific risk assessment and provide a logical starting point for screening lower-risk biological agents from inclusion in the full risk assessment.

B.2 THREAT ASSESSMENT (STEP 1B)

Next, we identify and evaluate possible adversaries and the threat they may pose to the facility (threat assessment). Where possible, facilities should use attributes of known adversaries; however, this is often impossible. An alternative is to create a set of notional adversaries whose attributes span the spectrum of plausible adversaries for the facility. We define a general set of notional adversaries here but, in order to increase the relevance, the risk assessor should modify these profiles based on data on the local threat environment. Facilities may also find that, in order to accurately analyze their threat environment, they need to create a new notional adversary with a specific set of defined attributes. The local law enforcement community is a good resource for this task. Reviewing the notional adversary descriptions that follow and completing the local law enforcement questionnaire (provided in Appendix A) are tools that can help initiate this discussion.

The notional adversary descriptions in this section are organized based on their level of authorized access. The descriptions define the assumption set used for each of the adversary attributes (motive, means, and opportunity). These adversary attributes (Equation B.6) provide a set of standardized criteria for doing the threat assessment for potential adversaries. These adversary descriptions are not intended to imply that all members of a group of individuals who have legitimate business at a facility will attempt to steal biological agents or other assets.

Equation B.6 Adversary Attributes

$$A = A_{mo}w_{mo} + A_{me}w_{me} + A_{op}w_{op}$$

where

A = Adversary attributes
A_{mo} = Adversary motive
A_{me} = Adversary means
A_{op} = Adversary opportunity
w_{mo} = Weight of A_{mo} criterion
w_{me} = Weight of A_{me} criterion
w_{op} = Weight of A_{op} criterion

Insiders are those individuals with authorized access. A facility may have only one type of insider (the Insider with Full Access), or it may have multiple types of insiders based on the level of access to the asset that might be stolen, such as Insider with Full Access, Insider with Building Access, and Insider with Site Access. There are some attributes that are assumed to apply to all Insider adversaries. An insider's motive may be due to disgruntlement, psychological imbalance, personal gain (in collusion), or the desire to commit a terrorist act. The malevolent insider would be expected to abort any theft attempt in order to avoid detection because, with authorized access, the insider can wait for a more opportune time to steal the asset. In general, the insider is an employee who has become malevolent and is not assumed to have paramilitary training.

The Insider with Full Access may be a laboratory worker or other individual who has unescorted access to the asset. Authorized access affords this person extensive knowledge of the facility and operating systems. It also provides this adversary with opportunity. Insiders with Full Access to biological materials are usually scientists and technicians who have a high level of technical training and sophistication. Thus, they often have all of the means at their disposal to successfully acquire and deploy a biological agent as a weapon. If a site implements access controls, then the risk assessor can modify the Insider with Full Access description to develop an appropriate set of assumptions for the additional insider types.

Outsiders are individuals without authorized access to the site. Information specific to the site and vicinity should be used to modify these notional descriptions or create additional ones.

The motive of the Terrorist Group adversary is to cause mass casualties, an economic crisis, or widespread fear, and/or it may seek to make a political statement. This adversary is assumed to be well funded and may be supported by a state, religious group, individuals, or even organized crime. Being well funded, the Terrorist Group is also well equipped and trained and able to rehearse an attack. The Terrorist Group is highly organized; it is violent and willing to die. It also has access to, and the skill to utilize, significant explosives and arms. As an Outsider, the Terrorist Group does not have access or specific opportunity.

A Single Terrorist adversary may be motivated to make a political statement, to express anger, to steal an agent in order to ultimately achieve personal goals, or to commit an act of bioterrorism. This adversary has less means than the Terrorist Group but is still well equipped and trained and able to rehearse. He or she may still be capable of killing or injuring a number of guards or other individuals. The Single Terrorist has the tools necessary to overcome most access control systems, and is willing to use violence and force. As an Outsider, the Single Terrorist does not have access or specific opportunity.

The Extremist Group adversaries are is motivated to make a political statement or protest against programs for ecological, political, economic, or other reasons. As a result, their objective may not be theft of a biological agent; instead, it may be to destroy property or release animals. However, their acts may inadvertently cause an escape of pathogens into the environment by releasing contaminated animals. The Extremist Group has means typically associated with sabotage, which may include hand tools, items for arson and other facility sabotage, and possibly handguns. It has general information about the facility but not specific information about the location of the assets or the facility's protection systems. All members of the group are assumed to be Outsiders; thus the Extremist Group does not have access or specific opportunity. Due to its presumed motive, this notional adversary is not typically included in risk assessments focused only on theft of pathogens and toxins for use in acts of bioterrorism.

The motivation of the Criminal is financial gain. The Criminal is a single adversary that may have weapons and hand tools. If organized crime is assessed to be a local problem, a Criminal Group adversary can be defined with appropriate means. The Criminal adversary is also presumed to be an Outsider without Access or specific opportunity. Criminal adversaries are also not usually included in risk assessments focused only on theft of pathogens and toxins for use in acts of bioterrorism.

The Competitive Rival seeks to gain competitive market advantage through theft or destruction of proprietary information or experimental materials. The Competitive Rival may include an invited colleague or other visitor who has the intent to steal intellectual property and/or acquire an asset. This adversary is assumed to have limited means but he or she may have opportunity with direct, supervised access (i.e., as an Escorted Insider). This adversary is also not relevant for bioterrorism risk assessments.

Vandals may operate individually or in groups. Their motivation is to cause a nuisance by damage or destruction. Tools include spray paint, knives, and hand tools, and may include guns for target shooters or hunters. They attack facilities in their vicinity and are not homicidal. They have no authorized access to the facility or opportunity to steal pathogens and toxins; they can be excluded from scenarios of theft of agents for use in bioterrorism.

A Colluding Terrorist Group is a combination of an Insider and an Outsider Terrorist Group. However, by involving more individuals, this notional adversary puts itself at risk of discovery.

1. A_{mo}: Motive characterizes why an adversary would steal a pathogen or toxin.

0	Adversary has no interest in biological agents.
1	Theft would be for personal gain (e.g., economic or revenge).
2	Adversary is interested in making a political statement.
3	Adversary seeks to conduct a small-scale bioterrorism incident.
4	Adversary intends to conduct a large-scale bioterrorism event, causing mass murder, mass hysteria, or devastating economic impact.

2. A_{me}: Means is a characterization of the adversary's technical skills, operational knowledge, and necessary tools required to conduct the scenario (i.e., theft and use of the biological agent).

0	Adversary has no means to execute the scenario.
1	Adversary has insufficient means to successfully execute the scenario.
2	Adversary has sufficient technical skills and tools but no operational knowledge.
3	Adversary has sufficient technical skills and tools and incomplete operational knowledge.
4	Adversary has extensive technical skills and operational knowledge and all of the necessary tools.

3. A_{op}: Opportunity characterizes whether an adversary could steal the biological agent covertly or must steal it overtly. This is based on the degree of the adversary's access to the asset.

0	Adversary has no legitimate access to the facility.
1	Adversary only has legitimate access to the facility site.
2	Adversary has unescorted access to the building with the asset and/or escorted access to the asset.
3	Adversary has occasional unescorted access to the asset.
4	Adversary has regular unescorted access to the asset and/or the opportunity to gain access to the asset 24 h a day, 7 d a week.

B.3 CREATING SCENARIOS (STEP 2A)

The next step is to screen assets that do not present sufficient risk to warrant a full assessment. In Chapter 2, we discussed eliminating biological agents from the full risk assessment that do not have at least a moderate baseline risk. Typically, many biological agents will be screened because they are nonpathogenic or have low malicious use risk (see Table B.1). Additionally, as discussed above, certain adversaries should be screened from certain scenarios, either because they have no interest in biological agents or because they have insufficient means.

Screening low-risk assets and adversaries that are incapable or uninterested in bioterrorism will help narrow the scope of the full risk assessment to a more manageable size while still leaving the risk assessor and risk manager with confidence in the results of the risk assessment.

Table B.2 shows the scenarios that are usually included in any biosecurity risk assessment. The risk assessor may, of course, include other scenarios deemed important from the facility's perspective. For an EMUR, it may be appropriate to assess scenarios involving the Insider, a Terrorist Group, and a Colluding Terrorist Group. Depending on the facility, a Single Terrorist may not have the means to steal an EMUR agent. For an HMUR, it may now be appropriate to include the scenario with a Single Terrorist. For an MMUR, the Terrorist Group (or Colluding Terrorist Group) may not have sufficient motive to target the specific facility being assessed for acquisition of these agents.

TABLE B.1
Examples of Biological Agents Screened from a Biosecurity Risk Assessment

Genus	Species	Biosafety Risk Group	Why Screened from Biosecurity Risk Assessment	Biosecurity Risk Group
		Bacteria		
Acinetobacter	*calcoaceticus*	2	Normal flora; opportunistic pathogen	Low risk
Staphylococcus	*aureus*	2	Ubiquitous in environment; opportunistic pathogen	Low risk
		Viruses		
Parvoviridae	Canine parvovirus type 2	2	Not a human pathogen; common; vaccine available	Nonpathogenic group
Picornavirus	Infectious encephalomyelitis	1	Not pathogenic to humans	Nonpathogenic group
		Fungi		
Candida	*albicans*	2	Ubiquitous; can cause mild human disease	Low risk
Penicillium	*Sp.*	1	Ubiquitous, rare reports of infection in immunocompromised hosts	Low risk

TABLE B.2
Scenarios in a Full Biosecurity Risk Assessment

Asset	Adversary	Action
EMUR	Insider	Theft of the pathogen or toxin
EMUR	Terrorist group	Theft of the pathogen or toxin
EMUR	Colluding terrorist group	Theft of the pathogen or toxin
HMUR	Insider	Theft of the pathogen or toxin
HMUR	Terrorist group	Theft of the pathogen or toxin
HMUR	Colluding terrorist group	Theft of the pathogen or toxin
HMUR	Single terrorist	Theft of the pathogen or toxin
MMUR	Insider	Theft of the pathogen or toxin
MMUR	Single terrorist	Theft of the pathogen or toxin

B.4 VULNERABILITY ASSESSMENT (STEP 2B)

In the context of a biosecurity risk assessment, this step entails reviewing the existing implementation of the biosecurity components: physical security, personnel security, MC&A, transport security, information security, and program management. The answers to the vulnerability questionnaires in Appendix A will be useful in rating the facility's implementation on a 0–4 scale for each biosecurity component.

Equation B.7 Vulnerability Assessment

$$V = V_{phy}w_{phy} + V_{per}w_{per} + V_{mca}w_{mca} + V_t w_t + V_i w_i + V_{pm}w_{pm}$$

where

V = Site vulnerability
V_{phy} = Physical security vulnerability
V_{per} = Personnel security vulnerability
V_{mca} = Material control & accountability vulnerability
V_t = Transport security vulnerability
V_i = Information security vulnerability
V_{pm} = Program management vulnerability
w_{phy} = Weight of V_{phy} criterion
w_{per} = Weight of V_{per} criterion
w_{mca} = Weight of V_{mca} criterion
w_t = Weight of V_t criterion
w_i = Weight of V_i criterion
w_{pm} = Weight of V_{pm} criterion

1. V_{phy}: This criterion captures the current implementation of physical security measures at the facility. There are a variety of physical security measures appropriate for bioscience institutions — facilities may be implementing each of these all of the time, most of the time, some of the time, infrequently, or not at all. For the purposes of this exercise, we recommend considering the manner in which the following physical security elements are currently implemented at the facility: controlled access to the laboratory, controlled access to the building, controlled access to the site, locked storage (e.g., refrigerators, freezers), adequate nighttime lighting at the building, unobstructed views of building entrances, and the ability to detect and assess unauthorized access (intrusion sensors, alarms, alarm assessment capability).

0	All of the physical security measures are implemented all or most of the time.
1	At least 50% of the physical security measures are implemented all or most of the time, and nearly all are implemented at least some of the time.
2	At least 50% of the physical security measures are implemented at least some of the time, and nearly all are implemented at least infrequently.
3	Some physical security measures are implemented.
4	No physical security measures are implemented.

2. V_{per}: This criterion captures the current implementation of personnel security measures at the facility. There are a variety of personnel security measures appropriate for bioscience institutions — facilities may be implementing each of these all of the time, most of the time, some of the time, infrequently, or not at all. For the purposes of this exercise, we recommend considering the manner in which the following personnel security elements are currently implemented at the facility: background screening of potential employees, escorts in restricted areas for nonauthorized personnel, records kept of employees with authorized access to restricted areas and key assignments, use of badges, use of photo identification badges, and out-briefings of departing employees.

0	All of the personnel security measures are implemented all or most of the time.
1	At least 50% of the personnel security measures are implemented all or most of the time, and nearly all are implemented at least some of the time.
2	At least 50% of the personnel security measures are implemented at least some of the time, and nearly all are implemented at least infrequently.
3	Some personnel security measures are implemented.
4	No personnel security measures are implemented.

3. V_{mca}: This criterion captures the current implementation of MC&A measures at the facility. There are a variety of MC&A measures appropriate for bioscience institutions — facilities may be implementing each of these all of the time, most of the time, some of the time, infrequently, or not at all. For the purposes of this exercise, we recommend considering the manner in which the following MC&A elements are currently implemented at the facility: an accountable individual is assigned to each MMUR, HMUR, or EMUR; pathogens or toxins that are not needed are destroyed; the direct laboratory supervisor is aware of all pathogens and toxins used or stored in the laboratory; a current inventory exists for all pathogens and toxins; and the inventory is reconciled through a physical inventory-taking process.

0	All of the MC&A measures are implemented all or most of the time.
1	At least 50% of the MC&A measures are implemented all or most of the time, and nearly all are implemented at least some of the time.
2	At least 50% of the MC&A measures are implemented at least some of the time, and nearly all are implemented at least infrequently.
3	Some MC&A measures are implemented.
4	No MC&A measures are implemented.

4. V_t: This criterion captures the current implementation of transport security measures at the facility. There are a variety of transport security measures appropriate for bioscience institutions — facilities may be implementing each of these all of the time, most of the time, some of the time, infrequently, or not at all. For the purposes of this exercise, we recommend considering the manner in which the following transport security elements are currently implemented at the facility: appropriate permissions are obtained prior to sharing pathogen or toxin samples, all transport is documented, secure storage is used in the shipping and receiving areas, the recipient laboratory is verified, timely shipping methods are used, successful receipt at the intended destination is verified, and internal transport controls are in place.

0	All of the transport security measures are implemented all or most of the time.
1	At least 50% of the transport security measures are implemented all or most of the time, and nearly all are implemented at least some of the time.
2	At least 50% of the transport security measures are implemented at least some of the time, and nearly all are implemented at least infrequently.
3	Some transport security measures are implemented.
4	No transport security measures are implemented.

5. V_i: This criterion captures the current implementation of information security measures at the facility. There are a variety of information security measures appropriate for bioscience institutions — facilities may be implementing each of these all of the time, most of the time, some of the time, infrequently, or not at all. For the purposes of this exercise, we recommend considering the manner in which the following information security elements are currently implemented at the facility: computer passwords are used, sensitive documentation is marked, sensitive documentation is protected from inadvertent public disclosure, a review-and-approval process is in place, computer and network security measures are used, and sensitive documentation is destroyed before being discarded.

0	All of the information security measures are implemented all or most of the time.
1	At least 50% of the information security measures are implemented all or most of the time, and nearly all are implemented at least some of the time.
2	At least 50% of the information security measures are implemented at least some of the time, and nearly all are implemented at least infrequently.
3	Some information security measures are implemented.
4	No information security measures are implemented.

6. V_{pm}: This criterion captures the current implementation of biosecurity program management at the facility. There are a variety of program management measures appropriate for bioscience institutions — facilities may be implementing each of these all of the time, most of the time, some of the time, infrequently, or not at all. For the purposes of this exercise, we recommend considering the manner in which the following program management elements are currently implemented at the facility: an individual is designated as a biosecurity officer, personnel receive appropriate biosecurity training, there is a written biosecurity manual, SOPs include biosecurity aspects, the institution provides sufficient resources to implement biosecurity, risk assessment is conducted annually, self-audits are conducted annually, exercises are conducted annually, biosecurity is integrated with biosafety, and there is attentiveness to personal issues of staff that could negatively impact biosecurity and biosafety.

0	All of the program management measures are implemented all or most of the time.
1	At least 50% of the program management measures are implemented all or most of the time, and nearly all are implemented at least some of the time.
2	At least 50% of the program management measures are implemented at least some of the time, and nearly all are implemented at least infrequently.
3	Some program management measures are implemented.
4	No program management measures are implemented.

B.5 EVALUATE THREAT POTENTIAL AND CONSEQUENCES OF SCENARIOS (STEP 3A)

Equation B.1 explains that the biosecurity risk is a function of threat potential and maximum credible consequences. To characterize the risk, for each scenario that has not been screened, the threat potential and maximum credible consequences must be analyzed. Threat potential is a function of agent task complexity, adversary attributes, and site vulnerability.

Equation B.8 Threat Potential

$$T = TCw_{TC} + Aw_A + Vw_V$$

where

TC = Agent task complexity
A = Adversary attributes
V = Site vulnerability
w_{TC} = Weight of TC criterion
w_A = Weight of A criterion
w_V = Weight of V criterion

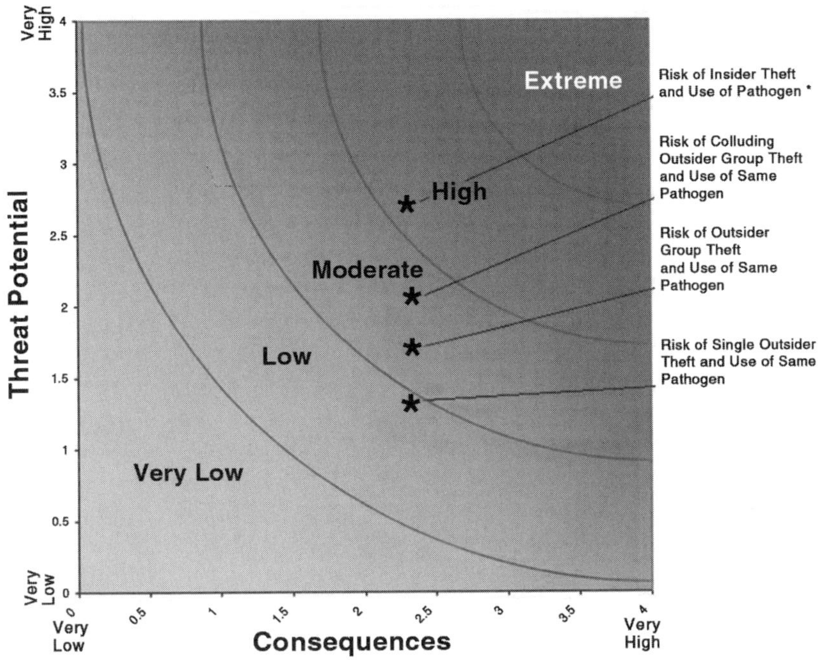

FIGURE B.1 **(See color insert following page 16.)** Hypothetical risk assessment results for one pathogen.

The consequences variable was defined in Equation B.4 as a function of the impacts on population, economic, psychological, and operational elements. Graphing threat potential vs. consequences allows for a visual representation of risk. Figure B.1 shows the results of a hypothetical risk assessment for four scenarios: four notional adversaries each executing the same action (theft and use) on the same pathogen. Graphical representation is instrumental in making the risk decision as to which risks are acceptable and which are unacceptable. In this set of results, a risk manager may decide to implement biosecurity to mitigate the risk of the insider stealing and using the pathogen while developing incident response plans for the other three (more acceptable) scenarios.

Appendix C

Biosecurity Plan Template

This appendix provides an example of a template that may be useful to bioscience facilities as they develop a biosecurity plan or manual. This template was originally created as an aid for institutions that need to comply with U.S. select agent regulations (42 CFR 73, 9 CFR 121, and 7 CFR 331), but we hope that the basic structure and principles will be applicable to any institution as it develops a biosecurity plan.

Generic text that may be appropriate to include in a facility's biosecurity plan is included in the following text. Guidance on facility-specific information that should be included is provided in italicized text.

C.1 INTRODUCTION

What is the goal of this plan? To whom does it apply? Indicate that it demonstrates compliance with any specific federal regulations, such as 42 CFR 73, 9 CFR 121, or 7 CFR 331, and that it describes the full spectrum of measures taken to achieve graded protection of Select Agents (which should be defined in this introduction as a term used in this plan to refer to all CFR-regulated pathogens and toxins) against theft and sabotage. Indicate whether a single approach is being taken to secure all Select Agents at the facility or whether Moderate-Risk and High-Risk agents are being addressed separately (while still complying with all federal regulations).

C.2 ROLES AND RESPONSIBILITIES

The roles and responsibilities included in this section are not all-inclusive but are intended to represent those functions related to implementation of the CFR requirements.

Note that multiple roles and responsibilities may be assigned to the same person, especially at institutions with small programs of work with dangerous biological agents.

C.2.1 Responsible Official

The Responsible Official (RO) is the one authorized to ensure that the requirements of the CFRs are met. These requirements include developing and implementing this Biosecurity Plan. The RO will review this plan annually and after any incident.

C.2.2 Alternate Responsible Official

The Alternate RO is an official authorized to act for the RO when the RO is unavailable.

C.2.3 Select Agent Supervisor

Select Agent Supervisors are individuals who are responsible for directing a project or program. Each Select Agent project or program is overseen by a Select Agent Supervisor who is responsible for the scientific and technical direction of that project or program, and who has task authority over individuals who have permission to use Select Agents. Select Agent Supervisors are responsible for:

- Adopting the Biosecurity Plan procedures and ensuring that all personnel within their charge who have access to Select Agents familiarize themselves with the contents of the plan and obtain biosecurity training annually
- Reporting Select Agent transfers, destruction, and inventory anomalies to the RO
- Requesting the RO to make Select Agent access authorization changes (see also Section 4.6.5)
- Providing the RO with any nonelectronic visitor logs upon request
- Requesting changes to personnel access authorization
- Providing the RO with an up-to-date Select Agent registration packet

C.2.4 Accountable Scientist

The accountable scientist, who may be a Select Agent Supervisor and/or a Principal Investigator, is responsible for Select Agent material control and accountability and Select Agent material transfers, as described in Section C.6 and Section C.7.

C.2.5 Security Force

If a security force is employed, what is the nature of its responsibilities?

C.2.6 Local Police

If applicable, what is the nature of the local police force responsibilities under a MOU?

C.2.7 Specialty Personnel

Specialty personnel may be employed by larger facilities. These may include Security Specialists who work in a Security Operations Center where an intrusion detection system is monitored, Physical Security Department Personnel, and Counterintelligence Personnel.

Roles and responsibilities for these personnel should be spelled out in this portion of the security plan.

C.2.8 Personnel Security

The Personnel Security Division is responsible for initiating and monitoring necessary background screening and often for evaluating the results.

C.2.9 Badge Office

Badge Office Personnel are responsible for issuing and managing badges for regular and visiting personnel.

C.2.10 Information and Network Security

Information and network security personnel include the following:

- The Chief Information Security Officer is responsible for network and information security policy for the facility.
- The Center and Division Information Technology Officers are responsible for ensuring that their respective network segments and information protection systems are implemented according to policy and that personnel are adequately trained on information and network security.
- System/Network Administrators are responsible for maintaining the system security, updating hardware and software, and responding to network intrusions.

C.2.11 Individuals with Select Agent Access Authorization

In addition to other duties individuals have, individuals with Select Agent access authorization are responsible for:

- Protecting Select Agents while in their physical possession
- Protecting information related to Select Agents, while in their physical possession, in the context of verbal or electronic communication, and when storing them
- Following all security-related procedures related to Select Agents, including those that apply to hosting and escorting procedures for visitors (see Chapter 5, Section 5.8 and Section 5.9)
- Reporting incidents and/or breaches in security to the appropriate Select Agent Supervisor and RO

C.3 BASIS FOR BIOSECURITY PROGRAM

C.3.1 Risk Assessment

This security plan reflects a risk management process in which assets and possible adversary actions (threats) are defined, and the resulting undesired events are evaluated based on their security risk. The risk assessment is an evaluation of the potential an adversary possesses to successfully execute an undesired event and the subsequent consequences. It establishes the set of risks a facility faces and presents them in ranked order so that facility management may decide which risks will be protected against or mitigated and which risks will not. The security plan is based on this defined security risk posture and demonstrates how the facility achieves protection and mitigation through a combination of security system design and incident response planning.

C.3.2 Graded Protection

Different assets require different levels of protection, accountability, and controls. The highest level of protection is given to the primary assets whose loss, theft, compromise, and/or unauthorized use will most seriously affect national security, and/or the health and safety of employees, the public, the environment, or mission (e.g., High-Risk pathogens). Slightly less protection is given to those secondary assets that may represent a Moderate Risk or that may assist an adversary in gaining access to, or diverting, a primary asset. Tertiary assets include operational assets and require somewhat less protection than the secondary assets. In this manner, the security system is designed to have graded levels, with the highest risk assets receiving the highest level of protection, and security increasing gradually as one moves physically closer to the asset.

What types of assets would be considered Primary, Secondary, and Tertiary at this facility?

C.3.3 Assets

C.3.3.1 Select Agents

Select Agents are those agents and toxins that have the potential to pose a severe threat to human, animal, or plant health, or to plant and animal products as defined by the CFRs.

Which Select Agents does the facility possess?

C.3.3.2 Sensitive Information

Sensitive information is information that is too sensitive to be released to the public or to anyone who does not have an official purpose that requires him or her to hear, view, or have possession of the information (i.e., a need to know). Sensitive information is protected from unauthorized access and from disclosure under the U.S. Freedom of Information Act (FOIA. See Section C.8 for details on protecting sensitive information.

Sensitive information includes information related to the Select Agents, security-related information, and human resources information specific to those individuals who work with Select Agents.

C.3.3.2.1 Select Agent Information

The following examples of sensitive information include, but are not limited to, the Select Agent records the RO is required to maintain:

- Select Agent information related to records described in the CFRs as:
  - A current list of all individuals with access to Select Agents
  - Training records for individuals with access to Select Agents
  - Select Agent inventory records (including source and characterization data as well as any anomalies)
  - Permits and transfer documents
  - Visitor logs for laboratories containing Select Agents
- Databases containing security and Select Agent information
- Documentation associated with experimental data or other data that have been restricted by the facility's review and approval process

C.3.3.2.2 Security-Related Information

The following examples of sensitive information include, but are not limited to, the security-related records the RO is required to maintain:

- Security information related to the records described in the CFRs:
  - Security records (e.g., transactions from automated access control systems, testing and maintenance of security systems, visitor logs)
  - Containment and security incident reports
  - Biosecurity Plan
- Details of facility description and blueprints especially as related to Limited and Exclusion Area designations and protection measures
- Details of vulnerabilities of those facilities that handle Select Agents and/or sensitive information
- Details of physical security (e.g., drawings and descriptions of security hardware and software systems)
- Details of computer systems and procedures
- Security procedures
- Badge design information
- Security system performance test results and audit results
- Incident reports and disciplinary actions
- Response force contracts and results of response force exercises

C.3.3.2.3 Human Resource Information

Human resource information includes all information about personnel who work with or have access to Select Agents. This information includes:

- Home contact information
- Listings of family members
- Financial information
- Background investigation results

C.3.3.3 Critical Operational Assets

Critical operational assets are those that may cause significant work delays or financial impact if destroyed or are directly involved in the security associated with High-Risk Agents.

Provide a list of the critical operational assets with a brief description of each.

C.3.4 Threat Definition

C.3.4.1 Insider

The insider threat category includes a single, nonviolent person with authorized access inside the facility. The insider is considered to be any person granted unescorted access to any portion of an Exclusion or Limited Area (see Section C.4.2 and Section C.4.3 for further details on these areas). The intent of a malevolent insider is to steal, destroy, or release a Moderate- or High-Risk agent, or to steal or destroy other high-consequence assets at [*facility name*] without detection. The insider would be expected to abort any theft attempt to avoid identification. Authorized access affords this person extensive knowledge of the facility and operating systems. The insider has the opportunity to choose the best time to commit a malevolent act.

C.3.4.2 Outsider

Outside adversaries can employ force, stealth, and deceit tactics to achieve their goals. Using force, the adversary makes no attempt to conceal acts or intention; the adversary simply overwhelms the system and personnel. Using stealth, the adversary attempts to enter the facility undetected to accomplish his or her goal. An adversary using deceit will attempt to accomplish his or her goal under the guise of authorized access through the use of forged credentials or other methods. Obviously, a sophisticated and well-trained adversary could employ a combination of all three tactics in order to steal, destroy, or release a defined asset. The outsider has access to only publicly available information, and may be equipped with hand tools, may be armed, and may resort to violence (but is not suicidal).

C.3.5 Protection Strategy

C.3.5.1 Insider Protection

Traditional physical protection measures, personnel security programs, strict escorting rules, and MC&A procedures are the basic elements of the security strategy for protection against a malevolent insider. Of increased importance, given the difficulty with pathogen accountability, is the reliance that must be placed on employees and others with access to the pathogens. Stand-off detection technologies do not exist for biological agents, and inventory control systems will not necessarily reveal when material has been stolen or diverted. Thus, the insider threat is a daunting problem for biological research laboratories.

It is very difficult for a physical security system to prevent the theft or diversion of microorganisms by insiders. Therefore, it is paramount that biological research facilities do everything possible to ensure that those who have access to dangerous pathogens and toxins are reliable and trustworthy.

It should be noted that foreign nationals cannot be investigated as thoroughly as U.S. citizens until the foreign national has resided in the U.S. for the number of years that the investigation will cover. Until this point in time is reached, foreign nationals holding positions requiring a background investigation will represent a relatively greater risk than U.S. citizens. Collusion is protected against in the same manner as any other insider threat.

C.3.5.2 Outsider Protection

The strategy to protect against an outsider is to detect unauthorized access, through likely avenues of approach, to the biosafety containment labs or other areas where critical assets are located. Detection must be done in a timely manner and response forces summoned. These response forces may be private security forces or local law enforcement (LE). When local LE is employed, it is important to have a MOU in place that outlines the conditions under which local LE will respond, the response time that may be expected, and the protocol to follow once LE arrives on site (due to possible biological containment issues).

The approach often used to achieve timely detection is to concentrate security upgrades at the physical locations where the pathogens or other critical assets are kept, and to control access to these locations.

C.4 PHYSICAL SECURITY

The physical security system limits access into defined security areas to authorized individuals with a valid need for access.

C.4.1 Property Protection Areas

A Property Protection Area is defined by the outermost perimeter of the facility. This security area is established to protect against damage, destruction, and theft of facility-owned property.

What establishes the Property Protection Area (e.g., a perimeter fence)? What, if any, credentials are required to access the Property Protection Area?

What areas of the facility are Property Protection Areas? What assets are within this area?

C.4.2 Limited Areas

A Limited Area is a secured area, residing within the Property Protection Area, with barriers that identify its boundaries and encompass the designated space. The perimeter of a building often defines the boundaries of a Limited Area.

What physical security measures are in place? What credentials are required to access the Limited Area?

What areas of the facility are Limited Areas? What assets are within these areas?

C.4.3 Exclusion Areas

An Exclusion Area, like a Limited Area, is a security area with barriers that identify its boundaries and encompass the designated space, further restricting access beyond the Limited Area. Laboratories or storage areas that contain Select Agents are often designated as Exclusion Areas.

What physical security measures are in place? What credentials are required to access the Exclusion Area?

What areas of the facility are Exclusion Areas? What assets are within this area?

C.4.4 Long-Term Select Agent Storage

Are there differences in which area select agents can be stored if they are in locked storage containers (e.g., freezers, refrigerators)? If so, include the locations here.

C.4.5 Security Operations

C.4.5.1 Access Hours

Does everyone have 24-h access, or do certain types of workers have access in different time zones, e.g., Monday–Friday, 6 a.m.–6 p.m.; Monday–Sunday, 6 a.m.–6 p.m.; or 24 h a day, 7 d a week.

C.4.5.2 Visitor Logs

In what rooms/areas are visitors required to sign logbooks? What information must be included? Does the escort also need to sign?

C.4.5.3 Vehicles

Who is authorized to park on site? Are there other parking restrictions, e.g., are private vehicles restricted from loading dock areas? Do personal cars require a parking sticker or placard? How is visitor parking handled?

C.4.5.4 Tailgating

"Tailgating" is the practice of one individual following another into an area that has been restricted with an electronic security device (e.g., a proximity card) without utilizing his or her own means for unlocking the door. Tailgating is prohibited into any Limited or Exclusion Area. The term *tailgating* is not used to describe the authorized entrance of individuals under escort, who follow additional escort/host procedures to ensure accountability.

C.4.5.5 Access Changes

When an individual is reassigned to an activity that does not require access to Select Agents, requires access to different Select Agents, or is initiating access to Select Agents, his or her access control authorizations must be updated. The RO reports any changes in Select Agent access to the CDC/APHIS/Select Agent Program (SAP). The RO will immediately notify CDC/APHIS/SAP when an individual's access to Select Agents is terminated; the RO must explain to CDC SAP/APHIS the reasons for terminating access. When an individual no longer needs access to a particular restricted area, these changes are also documented, and electronic access devices are updated.

C.4.5.6 Package Inspections

The CFRs require that all suspicious packages are inspected before they are brought into or removed from the area where select agents or toxins are used or stored.

Included under this section are details regarding the specifics of suspicious package inspections: Is the inspection conducted in Shipping & Receiving? At the entrance of the laboratory? What are they inspected for? Leaks, damage, etc? Who conducts package inspections?

Employees should be suspicious of any mail that

- Is unexpected or from someone unfamiliar to you
- Is addressed to someone no longer with your organization or is otherwise outdated
- Has no return address, or has one that cannot be verified as legitimate
- Is of unusual weight, given its size, or is lopsided or oddly shaped
- Is marked with restrictive endorsements, such as "Personal" or "Confidential"
- Has protruding wires, strange odors, or stains
- Shows a city or state in the postmark that does not match the return address

If you come in contact with any mail you consider suspicious, whenever you see an unattended or suspicious item on [*the site name*] properties, or if you receive a suspicious package, do not handle the item. Call one of the following numbers.

Call [*emergency phone number*] if the situation appears to be an emergency, or [*nonemergency phone number*] if the situation appears to be a nonemergency. Please be prepared to provide the location and description of the suspicious package.

While waiting for emergency response personnel to arrive, please follow the guidance below:

1. Do not handle the parcel or contents further.
2. Isolate the parcel or contents and move personnel from the immediate area.
3. Ensure that all personnel who have come into contact with the parcel or contents wash their hands thoroughly with soap and cold water.

C.5 PERSONNEL SECURITY

C.5.1 Position Risk Categories

C.5.1.1 Low Risk

Low-risk positions are those that involve duties with the potential for limited impact on the agency or program mission or on the integrity and efficiency of the services provided.

C.5.1.1.1 Background Investigation

What types of personnel screening does the facility use for people in this category?

C.5.1.1.2 Job Categories

All [*facility name*] employees, contractors, and working visitors who do not fall into the Moderate- or High-Risk categories are considered to hold Low-Risk positions. Personal and Casual Visitors are not given a risk designation.

What types of job categories at the facility are considered low risk?

C.5.1.2 Moderate Risk

Moderate-risk positions are those with duties that are of considerable importance to the agency mission, with significant program or delivery of service responsibilities. Moderate risk is the position risk level for the majority of positions associated with Select Agents.

C.5.1.2.1 Background Investigations

Those in moderate-risk positions typically receive a more comprehensive background investigation than those in low-risk positions. Sometimes this is limited to the additional requirement of the Department of Justice Risk Assessment if this category is limited to those who require access to Select Agents. These positions may also be subjected to a periodic reinvestigation.

What types of personnel screening does the facility use for people in this category?

C.5.1.2.2 Job Categories

What types of job categories at the facility are considered low risk?

C.5.1.3 High Risk

High-risk positions are those with duties that have a broad scope of responsibility and authority, which are especially critical to the agency or program mission.

C.5.1.3.1 Background Investigations

High-risk positions typically receive a more comprehensive background investigation and/or have a more restrictive authorization process. These positions may also be subjected to a periodic reinvestigation.

What types of personnel screening does the facility use for people in this category?

C.5.1.3.2 Job Categories

What types of job categories at the facility are considered low risk?

If an employee has access to classified information, he or she is considered to hold a high-risk position. Generally, those positions at the top of the executive ladder, and those in high-level positions of the security staff or IT staff who have access to the types of information that if released would make the facility vulnerable, are considered high risk.

C.5.2 Reinvestigations

The RO must request renewal of the CDC SAP/APHIS access approval every 5 years for as long as an individual needs access to Select Agents.

What position risk categories, if any, have background investigations that will be routinely repeated on a periodic basis? What is the period between investigations?

C.5.3 Access Limitations

C.5.3.1 Employees

Those individuals who require access to Select Agents must have CDC SAP/APHIS access authorization.

Are there any other access limitations in place for limited or exclusion areas? For example, must the background investigation be complete before an individual is allowed into an area without an escort? Must an individual meet any other requirements before being granted authorized access (e.g., yearly training, immunizations)?

C.5.3.2 Visitors

Visitors include personnel from universities, contractors, students, research fellows, visiting scientists, laboratory visitors, trades professionals, delivery personnel, etc., who, due to the duration of stay or nature of the work performed on site, are not provided with regular access to the facility. Visitors are escorted at all times in restricted (nonpublic) areas by an individual who has a complete and approved background investigation, access authorization, and a need to know. Visitors are expected to wear a visitor badge, sign all visitor logs, remain with their escorts, and follow all facility policies and procedures, including the surrender of prohibited articles while on site.

Note: Facility/security managers may permit a visitor to have unescorted access to restricted areas if the visitor is able to provide proof of an equivalent background investigation to that required of regular staff, including CDC SAP/APHIS authorization that has been processed by the facility's RO for access at the facility, if appropriate, and has legitimate business in these areas.

C.5.3.2.1 Host Responsibilities

Each visitor or group of visitors must have a host at the facility. The host must have a standard badge. The host is responsible for informing the visitor of the relevant policies and procedures, including access restrictions, prohibited articles, etc. The host may escort the visitor or arrange for a separate escort who also has a standard badge and authorized escort into the areas to be visited.

Include any department or individual that requires advance notice of expected visitor arrivals (e.g., Physical Security, receptionist, parking attendant, etc.), and what information is required (e.g., visitor name, arrival date, and duration of stay).

C.5.3.2.2 Escorting

An individual who has a standard badge and authorization to enter the areas to be visited must escort visitors.

Are there different ratios of visitors-to-escort that apply in different areas, e.g., administrative offices vs. laboratories? Are visitors allowed on site only during certain hours? Are there certain areas where an escort is unnecessary?

C.5.4 Foreign Nationals

Foreign nationals from countries that the U.S. Secretary of State has determined to be supporters of international terrorism will not be approved for escorted or unescorted access into Select Agent areas.

C.5.5 Badging

"Standard" badges will be issued to all employees and contractors. Visitors will be issued a distinct visitor badge. The name of the individual, name of facility, picture of the individual (for standard badges), and expiration date will be included on the badge.

This section should accurately describe what features are included in both the standard badge and the visitor badge, e.g., types of information, electronic access control (usually on standard badges), etc. This section should describe how long the badges are valid, e.g., 5 years for employees and contractors, or limited to the duration of stay for the visitor. It should describe any exceptions to wearing a badge (e.g., in laboratories or under other conditions where safety might be compromised). It should also describe the procedure an employee with a standard badge follows if his or her badge is forgotten, lost, or stolen.

C.6 MATERIAL CONTROL AND ACCOUNTABILITY

For the purposes of this section, "material" refers to repository stocks of Select Agents. Clinical samples and working stocks are not included.

C.6.1 Material Control

All Select Agent materials are associated with specific laboratories, which are identified by campus [*if there are multiple campuses of the facility*], building number, floor, and room number. When materials are stored, the container (such as a freezer, refrigerator, or vault) is locked to restrict access.

Laboratory inventory is checked on an as-needed basis to confirm that records correspond to actual materials. Any discrepancies are reported to the RO. The inventory review may be initiated by the laboratory staff, or by request from the RO.

Any change to the association of a material with a laboratory is considered a material "transfer" and is subject to the provisions of the Material Transfer Security section. Inventory records must be consistent with transfer operations.

Testing, diagnostic, and clinical samples are not controlled as part of the material inventory. Nevertheless, when isolates have been identified in clinical or diagnostic material as Select Agents, and those isolates are kept for future use, the isolates are added to material inventory as soon as they are stored.

Noninventory samples that may contain Select Agent material must be destroyed as soon as they are no longer needed. When inventoried material is destroyed, however, the inventory record is updated accordingly — the record is not deleted.

C.6.2 Accountability

Within each laboratory that uses or stores Select Agents, an accountable scientist maintains material inventory records, monitors the usage of materials, and oversees access to the materials. That accountable scientist is the contact person for the RO for any matters concerning the associated materials. The accountable scientist is usually a principal investigator or senior research scientist who normally works with the materials, and may be the Select Agent Supervisor or someone designated by the Select Agent Supervisor. If a laboratory works with more than one Select Agent, it may have separate accountable scientists for each agent. An accountable scientist may have a backup person, but otherwise the accountability is restricted to one individual. The accountable scientist ensures that the Select Agent Supervisor and RO are kept informed about changes to Select Agent inventories.

Other individuals within the same laboratory may have access to the areas where materials are stored and used, but the accountable scientist keeps the key to locked freezers or vaults where materials are stored. The accountable scientist may loan the key to others, but keeps a log of such usage. Others report what samples have been added or removed from storage so that the accountable scientist can keep the inventory records current.

C.6.3 Inventory Records

Each record of material inventory is kept in any one of a variety of forms as determined by the accountable scientist. It may be in the form of an electronic database (such as MS Access or SQL Server), spreadsheet files (such as MS Excel), or handwritten in logbooks or card files. The inventory record may also serve a research purpose and include additional information not required to meet the CFRs. Inventory records may also include non-Select Agent entries, but the Select Agent materials are identified as such to facilitate reporting. It is important that a single inventory be kept that satisfies both the biosecurity reporting requirements *and* any additional purpose.

Material inventory records are considered sensitive information, subject to the provisions of the Information Security section.

Each *record* in the material inventory refers to a distinguishable entity: either an individual item (such as a vial, ampoule, etc.) or collection of multiple items of the same Select Agent. If the latter, the record will further state the approximate quantity (such as mass, volume, number of items, etc.).

The information maintained about each inventory record (e.g., database *fields*) must include the following at a minimum, as required by 42 CFR 73.15 (b) and consistent with 7 CFR 331.14 (a) (4) and 9 CFR 121.15 (a) (4):

- Name and strain of the Select Agent
  - Source of the Select Agent:
    - How and when was the isolate acquired? Specific, private information about samples from human individuals is *not* required to meet 42 CFR 73 requirements, although it may be recorded for research purposes.
  - Location where stored:
    - The inventory record does *not* need to fully describe the location; for example, the rack/box/vial number may be specified explicitly, but the building/floor/room/freezer information may be the same for all, understood by the accountable scientist and omitted from the record.
- Amount
  - Approximate values are sufficient.
- Disposition
  - Date, recipient name, and approximate amount (for material transfer).
  - Date and approximate amount (for material destruction).
  - Explanation (for material lost, stolen, or otherwise unaccounted for).

42 CFR 73.15 (c) further mandates that a material *usage* log be kept for Select Agents. The material usage log may either be a separate table in a relational database, or usage could be recorded in a separate file. The usage information requires

- Name and strain of the Select Agent
- Who used it
- When removed and when returned to storage
- For toxins only: quantity removed and quantity returned

C.6.4 Reporting

The accountable scientist maintains inventory records so that the records reflect the daily current status of actual Select Agent inventories and usage. A copy of the inventory record and/or usage log is provided to the RO when requested.

The accountable scientist is also responsible for immediately reporting any of the following situations to the RO:

- Material is missing (whether lost, stolen, or otherwise unaccounted for)
- Material has been released outside of the biocontainment area accidentally or otherwise
- Inventory discrepancies: the physical inventory does not match the book inventory
- Anomalies potentially affecting inventory (e.g., if the key to a Select Agent freezer is missing)
- Any Select Agent samples, regardless of whether included in inventory or not, as soon as they are identified

The RO must immediately notify CDC or APHIS and appropriate federal, state, or local law enforcement agencies.

C.7 MATERIAL TRANSPORT SECURITY

Material transport requires authorization through the RO and CDC SAP/APHIS prior to conducting external transport. Materials transport must also have documentation and accountability maintained for Select Agents moving between Exclusion Areas during internal transport.

Personnel background screening requirements for individuals who have access to Select Agents packages (e.g., shipping, receiving, and internal delivery) are in accordance with the Personnel Security Section.

C.7.1 External Transport

External transfers consist of moving a Select Agent between a facility-registered entity and an external, authorized entity. External transfers require secure movement within a facility prior to relinquishing custody to a commercial carrier or courier company.

External transfers follow authorization and documentation procedures outlined in 42 CFR 73, 9 CFR 121, and 7 CFR 331. Therefore, when sending Select Agents, the accountable scientist, in conjunction with the RO, ensures that

- The sending and recipient RO possesses the appropriate certificate of registration to cover the transfer, the transfer meets the exemption requirements found in 42 CFR 73.6 (a), or the agent is being transferred from outside the U.S. and all import permits/requirements have been met.
- The requesting facility has a certificate of registration specific to the Select Agent to be transferred.
- Prior to shipping, the sender and RO for the recipient complete [*form*] and the sender submits the form to the proper agency for preshipping authorization.

When receiving Select Agent transport, the facility accountable scientist, in conjunction with the RO, ensures that

- The facility RO provides a completed copy of all necessary paperwork to the sender and the CDC SAP and/or APHIS within two business days of receipt.
- Any damaged packages containing Select Agents are immediately reported to the receiving RO. The recipient must also immediately report to the receiving RO when a shipment has not been received within 48 h of expected delivery time. In both cases, the RO must subsequently make an immediate report to the CDC SAP and/or APHIS as appropriate for the Select Agent involved.
- The recipient reports to the receiving RO, and the RO to the CDC SAP, within 5 d of material consumption or destruction in accordance with 42 CFR 73.21.

This facility also adheres to APHIS permit requirements found under the organisms and vectors regulation 9 CFR 122 and plant pest regulation 7 CFR 330, as well as all Department of Commerce (DOC) export permit requirements found in 15 CFR Part 742, 744, and 774.

C.7.2 Shipping and Receiving

This facility complies with all applicable transportation, shipping, packaging, and export laws related to Select Agents.

Which personnel are responsible for ensuring compliance with all appropriate regulations? Are there any specific facility shipping or receiving procedures?

C.7.3 Internal Transport

Internal Select Agent transport occurs as scientists and technicians exchange materials under study or add/remove Select Agents from the inventory through internal shipping and receiving processes. Any movement of Select Agent material into or out of a Select Agent–registered laboratory (i.e., all rooms managed by a single accountable scientist) is coordinated and authorized through the accountable scientist and must be documented in laboratory inventory records.

Are there any specific forms or procedures (such as Chain-of-Custody forms) for transporting Select Agents between registered laboratories within the facility?

C.8 INFORMATION AND NETWORK SECURITY

C.8.1 Information Security

C.8.1.1 Sensitive but Unclassified

Sensitive but Unclassified (SBU) is a designation that is applied by this facility to sensitive information that may be exempt from mandatory release to the public under the FOIA (5 U.S.C. 552). For information to be identified as SBU, it must fall within one or more of the FOIA exemption categories 2–9 (Exemption Category 1 is used for classified information).

Information related to Select Agents is generally unclassified, and if deemed sensitive, will be marked SBU.

Some of the information related to Select Agents that may be deemed sensitive includes

- *Databases and laboratory records associated with Select Agents, including, but not limited to, inventory databases and chain-of-custody records*
- *Select Agent transfer records*
- *Documentation associated with an experiment resulting in an unexpected result banned by 9 CFR Section 121.10*
- *Security documentation, personnel, and financial records (also often considered sensitive information)*

Note: If sensitive information at the facility is categorized as something other than SBU, change text to reflect facility's marking requirements.

C.8.1.2 Access to Sensitive Information

No person will be authorized access to sensitive information unless that person has been determined to need to know that particular information in order to achieve an authorized purpose.

C.8.1.3 Protection of Sensitive Information while in Use

Reasonable precautions will be taken to prevent access to sensitive information by persons who do not require the information to perform their jobs.

C.8.1.4 Communicating Sensitive Information

Sensitive information may be communicated in the following ways:

- From person to person in direct contact with one another
- Over a landline telephone
- Via first class, priority, or overnight mail, with no external markings that would indicate the material is sensitive
- Via fax machine when an authorized recipient is attending the machine
- By e-mail to and from a facility e-mail addresses [*e-mail address*] that reside completely within the facility network *(if the network is deemed sufficiently secure by the Chief Information Security Officer or other qualified individual)* or when the information is encrypted

C.8.1.5 Storage Rules for Sensitive Information

All sensitive information existing in hard copy or removable electronic media should be stored within a locked container in a Property Protection Area or within a Limited or Exclusion Area, an access-controlled electronic environment, or should be under the physical control of an authorized individual. When Limited or Exclusion Areas are not available, such as when an individual is traveling, a locked container within a locked room will suffice (e.g., locked briefcase or suitcase within a locked hotel room or vehicle).

C.8.1.6 Destruction of Sensitive Information

Sensitive information will be destroyed by shredding, or by any means available for classified information. Paper containing sensitive information will not be recycled.

Electronic or removable media will be physically damaged to the point of inoperability, via shredding, degaussing, melting, or other such methods before disposal.

C.8.1.7 Review and Approval Process

Review and approval of any information to be released to the public, through the medium of a publication, conference presentation, press release, Web site, or other form, will be conducted to ensure that sensitive information is not inadvertently released.

Specify who has the authority to approve the release of information.

C.8.1.8 Operational Security

All personal identification numbers (PIN), passwords, badges, lock combinations, keys, and key locations, or other access-related knowledge or devices, are controlled by the individual who has been given authority to receive this information or device. These items are not shared with any other individual.

C.8.2 Network Security

How is the facility's network protected? What encryption technique is used for electronic information?

C.8.2.1 Computer Processing

The following rules are applied when processing SBU information on the computer:

- Computers processing SBU information are facility-owned and protected as a part of the overall network security program.
- Computers employ password-locking screen savers to protect information on the screen and to secure the computer when it is not attended.

C.9 SAFETY

This facility follows the safety requirements as delineated for BSL 2, 3, and 4 laboratories, in accordance with the CDC/NIH publication *Biosafety in Microbiological and Biomedical Laboratories* (*BMBL*), fifth edition. Individual laboratories also develop and follow laboratory-specific safety manuals.

If alternate or additional safety requirements, guidelines, or manuals are followed, they should be specified here.

C.10 EMERGENCIES AND SECURITY INCIDENTS

Emergencies and security incidents that must be reported to the RO include the following.

C.10.1 Loss or Compromise of Access-Control Devices or Information

Access-control devices and information include keys, passwords, combinations, badges, card keys, and smart cards that provide access to Select Agents or their storage, use, or transport areas. Loss or compromise of these items or information must be reported immediately to Security and the RO. Loss or compromise of other forms of sensitive information must also be reported immediately to Security.

C.10.2 Unauthorized Persons

All personnel with unescorted access to a Limited or Exclusion Areas must be familiar with how to determine whether other individuals in the area are authorized to be there. They are to approach any visitor or other person who appears not to be authorized and/or observed to be conducting any suspicious activity in the area, as long as they appear nonthreatening, and challenge him or her by asking to see his or her badge. If the badge is a Visitor badge, and the individual is not under escort, the name of his or her escort and/or host must be obtained. If the individual or situation appears dangerous, or if the situation cannot be resolved, the authorized individual should contact the security manager immediately and leave the area. The RO must also be notified.

C.10.3 Loss, Theft, or Release of Select Agents

The RO must be notified upon discovery of inventory suspected of being lost, stolen, or misplaced, and any inventory records suspected of being altered.

Upon notification, the RO will investigate and notify the appropriate federal agency, if necessary, by completing CDC Form 0.1316 or APHIS Form 2043 as appropriate.

If the facility has spill response procedures, they should also be referenced here.

C.10.4 Safety Incidents involving Select Agents

Provide a reference to the appropriate facility response procedures.

C.10.5 Emergency Management

Provide a reference to the appropriate facility emergency management plans.

C.10.6 Suspicious Packages in Exclusion Areas

Provide a reference to the appropriate facility incident response plan.

C.11 TRAINING

Provide details on the facility's training requirements (both safety and security). What will the training cover? Who must complete the training? How often is retraining required? What are the penalties for failure to complete the necessary training? How are training records maintained?

C.12 INCIDENCE RESPONSE PLAN

See Sections 73.14, 121.14, and 331.14 regarding incident response planning.

"The incident response plan must fully describe the entity's response procedures for the theft, loss, or release of a select agent or toxin; inventory discrepancies; security breaches (including information systems); severe weather and other natural disasters; workplace violence; bomb threats; suspicious packages; and emergencies such as fire, gas leak, explosion, power outage, etc. The response procedures must account for hazards associated with the Select Agent and toxin and appropriate actions to contain such Select Agent or toxin.

The incident response plan must also contain the following information:

- The name and contact information (e.g., home and work) for the individual or entity (e.g., responsible official, alternate responsible officials, biosafety officer, etc.)
- The name and contact information for the building owner and/or manager, where applicable
- The name and contact information for tenant offices, where applicable
- The name and contact information for the physical security official for the building, where applicable
- Personnel roles and lines of authority and communication
- Planning and coordination with local emergency responders
- Procedures to be followed by employees performing rescue or medical duties
- Emergency medical treatment and first aid
- A list of personal protective and emergency equipment, and their locations
- Site security and control

- Procedures for emergency evacuation, including type of evacuation, exit route assignments, safe distances, and places of refuge
- Decontamination procedures

The plan must be reviewed annually and revised as necessary.

Drills or exercises must be conducted at least annually to test and evaluate the effectiveness of the plan. The plan must be reviewed and revised, as necessary, after any drill or exercise and after any incident."

Appendix D

Example Memorandum of Understanding with Local Law Enforcement

MEMORANDUM OF UNDERSTANDING BETWEEN INSTITUTION X

AND

LOCAL LAW ENFORCEMENT AGENCY MUTUAL ASSISTANCE AND INCIDENT RESPONSE AND RESOLUTION

I. Introduction
II. Purpose
III. Authority
IV. Response Guidelines
V. Resolution
VI. Funding and Liability
VII. Amendment and Term
VIII. Acknowledgment
Attachments

MEMORANDUM OF UNDERSTANDING BETWEEN INSTITUTION X

AND

LOCAL LAW ENFORCEMENT AGENCY MUTUAL ASSISTANCE AND INCIDENT RESPONSE AND RESOLUTION

I. INTRODUCTION

This Memorandum of Understanding (MOU) between Institution X and the Local Law Enforcement Agency (LE) sets forth understandings with regard to dealing with threats and criminal acts associated with theft or sabotage against Institution X. This MOU facilitates implementation of U.S. Presidential Decision Directives (PDD) 39 and 62, Homeland Security PD 5 (HSPD), the National Response Plan (NRP), and the National Incident Management System (NIMS).

The attorney general, through the Federal Bureau of Investigations (FBI), and in cooperation with other federal agencies engaged in activities to protect our national security, is responsible for activities of the law enforcement community to detect, prevent, and disrupt terrorist attacks against the U.S.

II. PURPOSE

The purpose of this MOU is to

1. Provide a clear basis for contingency response planning, coordination, and cooperation between LE and Institution X within the State of [State Y] to deal with terrorist threats and criminal acts against Institution X.
2. Facilitate effective communication and exchange of relevant information.
3. Assure timely, reliable, effective decision-making and response actions.

III. AUTHORITY

A. This MOU is established pursuant to PDDs 39 and 62, HSPD-5, the NRP, and NIMS.

B. Institution X Protective Forces are authorized under Act Z to carry firearms and make arrests without warrant in the protection of facilities and related activities; they are not authorized to investigate actual or alleged violations of Act Z.

C. LE is responsible for enforcement of state statutes, and city and county ordinances for the protection of property and civilians.

IV. RESPONSE GUIDELINES

A. Immediate Notification: When suspicious activity, a terrorist threat, or a criminal act occurs, a representative of Institution X will immediately notify LE of the incident and relay the status of the situation.

B. Command and Control: All actions by Institution X, LE, and other responding agencies will be coordinated through established emergency operations centers according to NIMS. On-scene officers will coordinate and establish appropriate interagency communication utilizing available resources or visual and verbal recognition (as directed by their respective commanders).

C. LE Response: LE will initiate a response and direct LE personnel to provide Institution X with requested law enforcement actions within their jurisdictional boundaries.

   1. LE will assist Institution X and other responding agencies by:
      a. Providing available resources to respond to the incident
      b. Establishing roadblocks on public roadways within its jurisdiction
      c. Participating in pursuit operations to stop fleeing persons who may pose a threat to the public if not apprehended
      d. Providing tactical officers to assist in recapture/recovery operations if or when an adversary force is in possession of sensitive assets and has established a stronghold position within jurisdictional boundaries for the purpose of perpetrating a criminal event
      e. Providing assistance in the command and control of responding law enforcement personnel to ensure an integrated level of communication is maintained between each agency
   2. The execution of any or all of these actions may expose LE to deadly force situations. LE policy for use of deadly force will be followed at all times by all LE officers.

D. Pursuit: During pursuit operations, command and control of all responding LE agencies will be based upon standard protocols and procedures adhered to by State Y LE agencies. Institution X protective force personnel will immediately relinquish lead control of the pursuit operation once LE joins the pursuit operation.

E. Institution X Protective Force Responsibilities: Where LE has initiated response to the incident, Institution X Protective Force will:

   1. Maintain responsibilities as the lead agency for hostile adversary situations involving interests within the confines of the facility property
   2. Ensure that all of their response resources remain available to support tactical operations
   3. Coordinate with other responding agencies for mutual assistance and response to the incident

4. Continue to cooperate with LE as required
5. Provide LE with assistance in the establishment of roadblocks required to restrict traffic flow to the area

V. RESOLUTION

A. Upon resolution, safety of personnel will be the first priority at the scene. Crime scene protection and evidence collection will be secondary. The crime scene and evidence involving Institution X materials will be secured and controlled by responding LE officers or federal officers if any are present at the scene.
B. Institution X will be available to assist LE with proper handling of stolen hazardous biological materials if this is necessary.
C. Custody of suspects will be maintained by LE unless federal LE agencies are on the scene. Suspects will be turned over to appropriate federal authorities depending upon the type of crime committed.
D. LE will coordinate with federal agencies to investigate the incident as per their established protocols.

VI. FUNDING AND LIABILITY

A. This MOU shall not be used to obligate funds or as a basis for the transfer of funds.
B. Nothing in this agreement shall be construed as inputting liability on any agency based solely on participation in this agreement.

VII. AMENDMENT AND TERMS

A. This MOU is subject to modifications required by Homeland Security's NRP.
B. This MOU shall become effective on the date of the last signature of involved parties. Any changes or amendments to this agreement are subject to approval of all involved parties. This MOU may be terminated by either party upon written notice.
C. No other agreements existing between Institution X and LE are affected, negated, or superseded by this agreement.
D. The procedures and protocols in this MOU will be exercised annually with all applicable LE agencies to ascertain effectiveness and to revise, update, or amend the MOU as required. The extent and complexity of LE participation in the annual exercise will be coordinated with LE.

VIII. ACKNOWLEDGMENT

The undersigned agree to this MOU and recognize each agency's status as a friendly force identified in the Institution X Site Security Plan.

______________________________	______________________________
LE Chief of Police	Institution X Representative
______________________________	______________________________
Date	Date

ATTACHMENTS

MEMORANDUM OF UNDERSTANDING BETWEEN INSTITUTION X

AND

LOCAL LAW ENFORCEMENT AGENCY MUTUAL ASSISTANCE AND INCIDENT RESPONSE AND RESOLUTION

OFFICIAL POINTS OF CONTACT

LE Police Department:

Chief of Police	Name	Phone number
Deputy Chief of Police	Name	Phone number

Institution X:

President	Name	Phone number
CEO	Name	Phone number
Biosafety Officer	Name	Phone number
Head of Security	Name	Phone number

Appendix E

SOP for Testing Access Control Systems

E.1 TEST PROCEDURES

These tests and procedures will check and provide assurance that the access control system components are functioning at a known performance level for authorizing the proper badges at the access-controlled doors. They will validate that alarms from those subcomponents give assurance that those components are protected with alarms or tampers, and that the components operate within the requirements of the statement of work. Three levels of testing can be performed on the access control and alarm system, depending on operational constraints imposed by the site or contractor.

1. Level 1 procedures will assure that electronic access-control locks and readers are in place at the doors. This level of testing will verify compliance with the Select Agent CFRs,[3] ensuring that limited access controls are in place.
2. Level 2 procedures will provide assurance that electronic access controls are in place at doors and that the capability to report an alarm if the door is forced open is in place. This will provide basic assurance of proper alarm reporting. It is not a full performance testing that would indicate that all hardware is functional nor that the conceptual design has been properly implemented.
3. Level 3 procedures will provide assurance that all hardware and critical software has been tested and meets the intent of the conceptual design. Checks are made on the documentation and training supplied by the installer.

Procedures are described for two door categories. The Category 1 Door is controlled by a proximity card reader or similar electronic credential reader and a magnetic lock. The doors are also equipped with a balanced magnetic switch (BMS) sensor and pneumatic door closer. The Category 2 Door is controlled by a proximity card reader or similar credential, PIN, or biometric device, and magnetic lock. The doors are also equipped with a BMS sensor and pneumatic door closer.

E.1.1 Level 1 Procedures

Note: Items in {} refer to the data sheet line (row) number.

Equipment

- Ballpoint pen or #2 pencil
- Badge and PIN that will allow access to all doors
- Badge and PIN that will not allow access to any doors (i.e., visitor)
- Data sheets for marking doorway hardware configurations

Field Test Procedure

1. Testers must have experience in identifying that the hardware mounted on each of the doors will form multiple teams. Tester will mark the data form with the location {2} and identification of the door and circle all visible hardware {5–19} on the outside door.
2. Tester optionally can check for the hardware {5–19} mounted on the inside of the doorway.
3. When the door is locked, the tester will pull and push (in an up/down and in/out motion) on door to ensure that the door does not open. Mark on {25} any movement that is noticed.
4. Category 1 door test
   - Tester will present a valid credential and verify {52} that the door opens. Tester does not need to go through the door.
   - Tester will present a nonvalid credential and verify {56} that the door does not open.
5. Category 2 door test
   - Tester will present a valid credential, enter an invalid PIN {53}, and verify that the door does not open.
   - Tester will present an invalid credential, enter a valid PIN {56}, and verify that the door does not open.
   - Tester will present a valid credential, enter a valid PIN {52}, and verify that the door opens. Tester does not need to go through the door.
6. Tester will verify that all information on the test data sheet is correct and continue to the next door.

This level of testing will only validate compliance with the CFRs using electronic access controls. It does not serve to validate proper installation of all hardware or programming. This level of testing is not recommended for validation and acceptance testing other than to verify the most basic operation of the access control system.

E.1.2 Level 2 Procedures

Note: Items in {} refer to the data sheet line (row) number.

Equipment

- Ballpoint pen or #2 pencil
- Badge and PIN that will allow access to all doors
- Badge and PIN that will not allow access to any doors (i.e., visitor)
- Badge that is not enrolled in the system
- Standard metal ruler at least 6 in. in length with markings to resolve measurements to 1/16th of an inch
- Data sheets for marking doorway hardware configurations
- Multiple two-way radios on the same frequency — one for the alarm/access control monitor and one for each field test team
- Strong large magnet to overcome the BMS magnetic field

Scripts for Level 2 Testing

All reference numbers in {} refer to the steps set forward in the level 2 testing procedures. Italicized words are actions the field tester and central alarm station (CAS) monitor will be doing to enter information on the data sheets.

CAS is defined as the person who is monitoring the alarm/access control monitoring station; if there are multiple stations, there should be one person at each station.

FT1 is defined as the field tester; if there are multiple groups, the numbers will be different. Only one set of scripts is indicated; the script for the field tester will be the same for each test group, but the CAS monitor ***must*** identify which field test group they are acknowledging, i.e., "FT1, I received your alarm. FT2, your door is reset."

Level 2 Script

FT1: CAS, I'm at Door ### ready to test locked door.

CAS: FT1, start locked door test.

FT1: (after push/pull [3] test), CAS, this is FT1; did you receive any alarms?

CAS: FT1, no alarms received. *Mark line? on data sheet "no alarm," if an alarm occurred, mark line ? that alarm received; if known, mark reason door alarmed.*

1. Testers must have experience in identifying that the hardware mounted on each of the doors will form multiple teams.
2. Tester will mark the data form with the location {2} and identification of the door and circle all visible hardware {5–19} on the outside door.
3. From the outside of the door, when the door is locked, the tester will pull and push (in an up/down and in/out motion) on the door

to ensure that the door does not open or provide for a visible gap between the door and the door jamb anywhere from the top to the bottom of the doorway. A visible gap is defined as a large enough opening that a small object, such as a #2 pencil or standard ballpoint pen, could fit through without applying undue force on the object. Mark on {25} any movement that is noticed. Mark after verifying with the CAS the alarm/no alarm {25} status during test.

4. Category 1 door test (if category 2 door, skip this step)
   - Tester will present a nonvalid credential and verify {56} that the door does not open.
   - Tester will present a valid credential and verify {52} that the door opens.
5. Category 2 door test (if category 1 door skip this step)
   - Tester will present a valid credential, enter an invalid PIN {53}, and verify that the door does not open.
   - Tester will present an invalid credential, enter a valid PIN {56}, and verify that the door does not open.
   - Tester will present a valid credential, enter a valid PIN {52}, and verify that the door opens.
6. Tester uses valid credential and PIN {52} to check for the hardware mounted on the inside of the doorway {5–19}, but not on any containment lab inner airlock doors.
7. When the door is locked, the tester will pull and push (in an up/down and in/out motion) on the door to ensure that the door does not open or provide for a visible gap between the door and the door jamb anywhere from the top to the bottom of the doorway. Mark on {26} any movement that is noticed. Mark after verifying with the CAS the alarm/no alarm {26} status during test.
8. Category 1 door test (if category 2 door, skip this step)
   - Tester will present an invalid credential and verify {56} that door does not open.
   - Tester will present a valid credential and verify {52} that door opens.
9. Category 2 door test (if category 1 door, skip this step)
   - Tester will present a valid credential, enter an invalid PIN {53}, and verify that the door does not open.
   - Tester will present an invalid credential, enter a valid PIN {56}, and verify that the door does not open.
   - Tester will present a valid credential, enter a valid PIN {52}, and verify that the door opens.
10. Tester will verify that all information on the test data sheet is correct and continue to the next door.

E.1.3 Level 3 Procedures

1. The door-monitoring sensor will detect movement of the door before that movement reaches 1 in.
2. It will verify that movement of the door of less than a half inch will not cause a false alarm indication from the device.
3. The access controls for the door will be checked for operation with the following.
   a. Authorized cards
   b. Unauthorized cards
   c. Wrong PIN
   d. Door open too long
   e. Door forced
4. Access controls will check all tampers associated with the door hardware and signal junction boxes with the following.
   a. Door position monitoring device
   b. Signal junction boxes
   c. Badge readers
5. The alarm monitoring station will be checked for reporting of all alarms and tampers from the access-controlled doors and signal junction boxes associated with protecting the site.
   a. Are the proper procedures posted?
   b. If graphical maps are used, are they accurate?
   c. What is the approximate time for alarm generation to alarm reporting?
   d. Are the alarm text indications accurate?

Other information that will be requested concerns the operation and maintenance of the system. This information will focus on the documentation of the system and its components:

1. How to operate and maintain the system
2. A schedule of training classes, if any were provided
3. Documentation of training of the operators and system administrators
4. Complete wiring and termination drawings of the system
5. Configuration information indicating all of the current setups (delays, door open times) and digital addresses associated with each access control component

Although these procedures do not attempt to check every possible configuration of the system for all circumstances, they do check the hardware and its functions to an acceptance level at a known performance level.

The procedure for each set of door hardware may vary depending on the configuration of the door and will use the attached data collection performance sheets to drive each test. The basic flow for the hardware tests are outlined below:

Test Performance Measure for BMS Door

1. Test door for movement while door is in the closed state. An alarm shall not be generated with door movement of up to a half inch. This will involve pulling and pushing the door against the locking mechanism. Record any BMS alarms generated from movement of the door. Test door for approximately 15 seconds using strong pull and push motion against the door. Repeat the test 10 times. Door will pass test if no alarms are generated.
2. Unlock the door-locking mechanism and measure door movement. An alarm shall be generated by the time the leading edge of the door has moved 1 in. from the closed position against the door jamb. The measurement is made beginning with the leading edge of the door against the door jamb in the closed position. From this position, open the door slowly until an alarm occurs. Measure the distance that the leading edge of the door has moved. The BMS passes if the movement distance is 1 in. or less. Repeat the test 10 times. Any failure to detect movement within 1 in. will result in a failure of the test.

Test Performance Measures for Access Control System

1. Authorized badge and PIN should release the electronic lock and system should record a transaction (badge, date/time stamp).
2. Unauthorized badge and/or PIN should not release the electronic lock and system should record attempted entry (badge, date/time stamp).
3. Generate a door open alarm and time to see that the alarm is received by the alarm station within 1 sec.
4. Door open time will be tested by using an authorized badge and PIN. The time from the entry of the last PIN to the time that the door is unlocked will be timed. This should not exceed 3 sec. This test should be repeated 10 times.

E.2.1 Example Test Data Sheet

Door Location						
Door Category	1	2	Other		Comments:	
Access Control Equipment located at Door						
Access Controller Junction Box ##		______				
BMS	Type				Sealed	Not Sealed
Reader In	Prox	PIN		Reader #		
Reader Out	Prox	PIN		Reader #		
RTE Out	IR	Button	Crash Bar	Other		

Emergency Pull Station					
Audio Sounder	Model				
Mag Lock	Model		600lb	1200lb	Other
Interlocked	Yes	No			
Door Information					
Mechanical Hardware	Inside	Outside			
Doors	Single	Double			
Makeup	Steel	Other			
Window	Small	Large	Meshed	Size	
Location	Interior	Exterior			

Comments:

BMS Tests

CAS MSG	Door Closed 15 Second Test				1st	2nd
	Outside	Alarm	No Alarm	Measured Door Movement		
	Inside	Alarm	No Alarm	Measured Door Movement		

CAS MSG	Door Open Alarm					1st	2nd
	Test 1	Alarm	No Alarm	Measured Door Movement			
	Test 2	Alarm	No Alarm	Measured Door Movement			
	Test 3	Alarm	No Alarm	Measured Door Movement			
	Test 4	Alarm	No Alarm	Measured Door Movement			
	Test 5	Alarm	No Alarm	Measured Door Movement			
	Test 6	Alarm	No Alarm	Measured Door Movement			
	Test 7	Alarm	No Alarm	Measured Door Movement			
	Test 8	Alarm	No Alarm	Measured Door Movement			

	Test 9	Alarm	No Alarm	Measured Door Movement			
	Test 10	Alarm	No Alarm	Measured Door Movement			

	BMS Alarm Large Magnet		No Alarm	Alarm
	BMS Tamper Small Magnet Door Open		No Alarm	Alarm
	BMS Tamper small Magnet door closed		No Alarm	Alarm
	BMS Case Open	N/A Sealed	No Alarm	Alarm
	BMS Junction Box	N/A no Swt	No Alarm	Alarm

Interlock Check (Anti-passback)

N/A	Outer Door Held Open	Inner Door	Entry	No Entry	Alarmed	No Alarm
N/A	Inner Door Held Open	Outer Door	Entry	No Entry	Alarmed	No Alarm

CAS MSG	**Outside Reader Tests** (If Cat 1, PIN not necessary) [If configured PIN only, valid credential is the PIN]						
	Configured for	Badge	PIN	Entry	Exit	Try Count=	
	Authorized Badge, Good Pin		Entry	Entry Time	No Entry	Alarm	No Alarm
	Authorized Badge, bad Pin		Entry	Entry Time	No Entry	Alarm	No Alarm
	Authorized PIN, no badge		Entry	Entry Time	No Entry	Alarm	No Alarm
	Unauthorized PIN, no badge		Entry	Entry Time	No Entry	Alarm	No Alarm
	Unauthorized Badge, good PIN		Entry	Entry Time	No Entry	Alarm	No Alarm

Reader Housing Tamper Yes No

Inside Reader Test (If Cat 1, PIN not necessary) [If configured PIN only, valid credential is the PIN]

	Configured for	Badge	PIN	Entry	Exit	Try Count=	
	Authorized Badge, Good Pin		Entry	Entry Time	No Entry	Alarm	No Alarm
	Authorized Badge Bad Pin		Entry	Entry Time	No Entry	Alarm	No Alarm
	Authorized Pin, no badge		Entry	Entry Time	No Entry	Alarm	No Alarm
	Unauthorized Pin, no badge		Entry	Entry Time	No Entry	Alarm	No Alarm
	Unauthorized Badge, Good Pin		Entry	Entry Time	No Entry	Alarm	No Alarm
	Reader Housing Tamper		Yes	No			

Reader Controller Time Tests Successful Door Open

		< 5 secs	> 5 secs
Test 1	Approximate Open Time		
Test 2	Approximate Open Time		
Test 3	Approximate Open Time		
Test 4	Approximate Open Time		
Test 5	Approximate Open Time		

CAS MSG	**Alarm Report Time (Door Forced)**			
	Test 1	Approximate Alarm Time		
	Test 2	Approximate Alarm Time		
	Test 3	Approximate Alarm Time		
	Test 4	Approximate Alarm Time		
	Test 5	Approximate Alarm Time		
	Test 6	Approximate Alarm Time		
	Test 7	Approximate Alarm Time		
	Test 8	Approximate Alarm Time		
	Test 9	Approximate Alarm Time		
	Test 10	Approximate Alarm Time		
	Badge Reader Controller Tamper			
	Intelligent Controller Tamper			

Other Door Hardware Tests		Single Leaf Door		Second Leaf	
Test	RTE	Opened	No Open	Opened	No Open
	Crash Bar	Opened	No Open	Opened	No Open
	Emergency Pull Station	Opened	No Open	Opened	No Open
	Other	Opened	No Open	Opened	No Open
	Audio Sounder	Yes	No	Yes	No
Comments					

APPENDIX F

Biosecurity Guidance and Regulations

U.S. LEGISLATION, REGULATIONS, AND GUIDANCE

Public Health Security and Bioterrorism Preparedness and Response Act of 2002, Public Law 107-188 June 12, 2002

http://www.fda.gov/oc/bioterrorism/PL107-188.html

Uniting and Strengthening America by Providing Appropriate Tools Required to Intercept and Obstruct Terrorism (USA PATRIOT Act), Public Law 107-54 October 26, 2001

https://www.biosecurity.sandia.gov/legislation

Antiterrorism and Effective Death Penalty Act, Law 104-132 April 24, 1996

https://www.biosecurity.sandia.gov/legislation

42 CFR 73 Department of Health and Human Services: Possession, Use, and Transfer of Select Agents and Toxins; Final Rule

http://www.gpoaccess.gov/cfr/index.html

9 CFR 121 Department of Agriculture: Agricultural Bioterrorism Protection Act of 2002; Possession, Use, and Transfer of Biological Agents; Final Rule

http://www.gpoaccess.gov/cfr/index.html

7 CFR 331 Department of Agriculture: Agricultural Bioterrorism Protection Act of 2002; Possession, Use, and Transfer of Biological Agents; Final Rule

http://www.gpoaccess.gov/cfr/index.html

49 CFR Parts 100-185 Department of Transportation Hazardous Materials Regulations

http://www.gpoaccess.gov/cfr/index.html

Biosafety in Microbiology and Biomedical Laboratories

http://www.cdc.gov/od/ohs/biosfty/bmbl4/bmbl4toc.htm

OTHER COUNTRIES' LEGISLATION, REGULATIONS, AND GUIDANCE

Canada: Laboratory Biosafety Guidelines

http://www.phac-aspc.gc.ca/ols-bsl/lbg-ldmbl/index.html

China: Regulations on Biosafety Management of Pathogenic Microorganism Laboratories

http://www.biosecuritycodes.org/leg.htm

Japan: Ministerial Notice on Laboratory Safeguards

http://www.biosecuritycodes.org/leg.htm

Singapore: The Biological Agents and Toxins Act of 2005

http://www.biosecuritycodes.org/leg.htm

United Kingdom: The Anti-Terrorism, Crime, and Security Act of 2001

http://www.opsi.gov.uk/ACTS/acts2001/20010024.htm

INTERNATIONAL REGULATIONS AND GUIDANCE

Biological Weapons Convention

http://www.opbw.org/

The 1925 Geneva Protocol/ Protocol for the Prohibition of the Use in War of Asphyxiating, Poisonous, or Other Gases, and of Bacteriological Methods of Warfare

http://www.fas.org/nuke/control/geneva/text/geneva1.htm

OECD Biosecurity Resources

http://www.biosecuritycodes.org/

UN Security Council Resolution 1540

http://disarmament2.un.org/Committee1540/

UN Model Regulations on the Transport of Dangerous Goods, *The Orange Book*

http://www.unece.org/trans/danger/publi/unrec/rev14/14files_e.html

World Health Organization: Laboratory Biosafety Manual, 3rd edition

http://www.who.int/csr/resources/publications/biosafety/WHO_CDS_CSR_LYO_2004_11/en/

World Health Organization: Biorisk Management Laboratory Biosecurity Guidance

http://www.who.int/csr/resources/publications/biosafety/WHO_CDS_EPR_2006_6/en/index.html

World Health Organization: Guidance on Regulations for the Transport of Infectious Substances

http://www.who.int/csr/resources/publications/biosafety/WHO_CDS_CSR_LYO_2005_22/en/index.html

OTHER LABORATORY BIOSECURITY RESOURCES

American Biological Safety Association

http://www.absa.org

Asia-Pacific Biosafety Association

http://www.a-pba.org

European Biosafety Association

www.ebsaweb.eu

Sandia National Laboratories' International Biological Threat Reduction Program

http://www.biosecurity.sandia.gov

U.S. Department of State's Biosecurity Engagement Program

http://www.BEPstate.net

U.S. National Academies

http://www7.nationalacademies.org/biosecurity/

U.S. National Science Advisory Board on Biosecurity

http://www.biosecurityboard.gov/

REFERENCES

1. Linkov, I., Satterstrom, K., Kiker, G., Bridges, T., Ferguson, E., and. Nelson, J., Multi-criteria decision analysis and homeland security applications, *Proceedings of the 2005 Conference, Working Together: Research and Development (R&D) Partnerships in Homeland Security*, Boston, MA, April 27–28, 2005.
2. http://mat.gsia.cmu.edu/mstc/multiple/node4.html.
3. 42 CFR 73, 9 CFR 121, and 7 CFR 331.

Index

C

D

E

F

G

H

I

J

K

L

M

N

O

P

Q

R

S

T

U

V

W

Y